T0349841

# SOIL
# MECHANICS

# SOIL MECHANICS

**Hongjian Liao**
Xi'an Jiaotong University, China

**Hangzhou Li**
Xi'an Jiaotong University, China

**Zongyuan Ma**
Xi'an University of Technology, China

NEW JERSEY • LONDON • SINGAPORE • BEIJING • SHANGHAI • HONG KONG • TAIPEI • CHENNAI • TOKYO

*Published by*

World Scientific Publishing Co. Pte. Ltd.
5 Toh Tuck Link, Singapore 596224
*USA office:* 27 Warren Street, Suite 401-402, Hackensack, NJ 07601
*UK office:* 57 Shelton Street, Covent Garden, London WC2H 9HE

Library of Congress Control Number: 2020941723

**British Library Cataloguing-in-Publication Data**
A catalogue record for this book is available from the British Library.

《土力学》
Originally published in English by Xi'an Jiaotong University Press


**SOIL MECHANICS**



ISBN 978-981-3238-50-3 (hardcover)
ISBN 978-981-3238-51-0 (ebook for institutions)
ISBN 978-981-3238-52-7 (ebook for individuals)

For any available supplementary material, please visit
https://www.worldscientific.com/worldscibooks/10.1142/10945#t=suppl

Desk Editors: Balasubramanian Shanmugam/Yu Shan Tay

Typeset by Stallion Press
Email: enquiries@stallionpress.com

Printed in Singapore

# Preface

The idea for this book was conceived when I was invited as a visiting scholar and attended some courses on soil mechanics at Kyoto University in Japan from 2006 to 2008. In 2011, I was once again invited as a JSPS visiting scholar to conduct research in geotechnical engineering as well as to attend a short-term course on soil mechanics at Kyoto University. During the Spring of 2012, I carefully compiled the materials presented in that course into a handbook.

Once I started teaching, I began to use bilingual teaching mode in my courses for undergraduate students at the Xi'an Jiaotong University in China. My handbook was recommended as a reference material for students. As years passed, I have tried to include the suggestions that were put forth by those who attended these courses. I also realized the need to compile a textbook in English. Fortunately, with the support of the "12th Five-Year Plan of Xi'an Jiaotong University" project and as a result of meticulous preparation, the first version of the book was published by the Xi'an Jiaotong University Press in 2015. The book is already widely used in a few universities in China as both a textbook as well as a reference book.

For complementing the quality of the currently available undergraduate courses in civil engineering as well as for attracting international readership, World Scientific Publishing Co. Pte. Ltd. has undertaken the task of publishing my book on the recommendation of the Xi'an Jiaotong University Press, which I consider to be a great honor. Therefore, the contents of this book were carefully revised and reorganized for better understanding for students.

Great attention has been paid to explain the basic theory, principles, and skills in soil mechanics for improving the students' understanding. All codes and standards included in this book are those that are currently being followed in China. At the beginning of each chapter, a guideline is provided to help students to systematically and comprehensively understand the topics. The English–Chinese–Japanese translation of the frequently used terminologies in soil mechanics have been provided in Appendix A for easy understanding for the students reading this book.

There are eight chapters in this book that deal with the following topics: basic characteristics and engineering classification of soils, permeability of soil and seepage force, stress distribution in soil, compression and consolidation of soil, shear strength, bearing capacity, stability of slopes and lateral earth pressure and retaining walls. Exercises and references are provided at the end of each chapter, with the answers provided at the end of the book. This book is especially suitable for undergraduate and masters' students and geotechnical engineering staff and can also be used as a reference book.

I would like to thank all the people who have supported, read, and provided suggestions on this book over the years.

The original version of *Soil Mechanics* was published by the Xi'an Jiaotong University Press in 2015 under my general editorship. The contributing authors for that edition included Prof. Lijun Su, Assoc. Prof. Hangzhou Li, and Assoc. Prof. Zhenghua Xiao. Their contributions to the various chapters were as follows: Chapters 1 and 2 and Appendices were written by Prof. Hongjian Liao (Chief Editor); Chapters 3 and 4 were written by Zhenghua Xiao, Hongjian Liao, and Hangzhou Li; Chapter 5 was written by Hongjian Liao and Hangzhou Li; Chapter 6 was written by Hangzhou Li; and Chapters 7 and 8 were written by Lijun Su. My sincere thanks to all of them.

Some of the contents in this book published by World Scientific Publishing Co. Pte. Ltd. have been revised carefully and systematically. The contributing authors for this book are Associate Professor Hangzhou Li and Associate Professor Zongyuan Ma. Their contributions to the various chapters are as follows: Introduction, Chapters 1–3, and Appendices were written by Professor Hongjian

Liao (Chief Editor); Chapters 4–6 were written by Associate Professor Hangzhou Li; and Chapters 7 and 8 were written by Associate Professor Zongyuan Ma. My thanks to them as well.

I am grateful to my Ph.D. student Yuqi He for his assistance in amending some of the contents and Ms. Yuan Bao of Xi'an Jiaotong University Press for her kindly help in publishing this book. I am also grateful to all the authors whose books were referenced.

I particularly wish to express my sincere thanks to the Xi'an Jiaotong University Press for recommending and supporting this book.

**Liao Hongjian**
Xi'an, China
December 2019

# About the Authors

**Liao Hongjian**, Ph.D., is currently a Professor and the Head of Department of Civil Engineering at Xi'an Jiaotong University. She graduated with a B.Eng. degree in 1985 and was awarded an M.Sc. degree in 1987 from the Department of Civil Engineering, Xi'an University of Architecture and Technology. She received a Ph.D. in Geotechnical Engineering from the Department of Civil Engineering, Tokai University, Japan in 1996. She is a member of the International Society for Soil Mechanics and Geotechnical Engineering, International Association for Engineering Geology and the Environment, and International Society for Rock Mechanics. She is also a Director of the Chinese Society for Rock Mechanics and Engineering, a Standing Director of the Environmental Geotechnical branch of the Chinese Society for Rock Mechanics and Engineering, a Deputy Chairman of the Specialized Committee of Soil Mechanics Teaching of the Soil Mechanics and Geotechnical Engineering branch of the China Civil Engineering Society, a Director of the Specialized Committee of Soil and Rock Mechanics branch of the Chinese Society of Theoretical and Applied Mechanics, and a Deputy Chairman of Shaanxi Society for Rock–Soil Mechanics and Engineering. She is an editorial board member of the *Chinese Journal of Rock and Soil Mechanics* and *Chinese Journal of Underground Space and Engineering*. She has taught soil mechanics for more than 30 years, including courses on soil mechanics, engineering geology, and advanced soil mechanics.

**Li Hangzhou**, Ph.D., is an Associate Professor at Xi'an Jiaotong University. His research interests include constitutive model of geomaterials and stability of underground excavations. He received a B.Eng. degree from China University of Mining and Technology in 2000, an M.Sc. degree from Xi'an University and Science and Technology in 2003, and a Ph.D. degree from Xi'an Jiaotong University in 2007. He worked as a research fellow in Nanyang Technological University, Singapore, from 2008 to 2010. His research projects include funding from the National Natural Science Foundation of China and Shaanxi Province Government. He has published more than 40 papers in academic journals and academic conferences. He is a member of the Chinese Society for Rock Mechanics and Engineering and Shaanxi Society for Rock–Soil Mechanics and Engineering.

**Ma Zongyuan**, Ph.D., is an Associate Professor at the Xi'an University of Technology. He specializes in computational geomechanics and earthquake engineering. His work focuses on slope stability and seismic analysis of geotechnical engineering. He received his M.Sc. degree from Chang'an University in 2006. He obtained his Ph.D. from Xi'an Jiaotong University (solid mechanics) in 2011. He joined the Xi'an University of Technology as a lecturer at the Institute of Geotechnical Engineering. He worked as a research fellow at the Colorado School of Mines, USA, from 2018 to 2019. He is a member of the American Society of Civil Engineers (ASCE) and International Society for Soil Mechanics and Geotechnical Engineering (ISSMG). He is the reviewer of several international peer-reviewed journals, such as *Journal of Engineering Mechanics ASCE*, *International Journal of Geomechanics ASCE*, *Marine Georesources & Geotechnology.* Dr. Ma has been the author and co-author of over 10 research publications in international peer-reviewed journals.

# Contents

*Preface* v

*About the Authors* ix

*List of Symbols* xv

*Introduction* xix

**1. Basic Characteristics and Engineering Classification of Soils** **1**

1.1 Introduction . . . . . . . . . . . . . . . . . . . . 2
1.2 Soil Composition and Phase Relationships . . . . . . 3
1.3 Soil Fabric . . . . . . . . . . . . . . . . . . . . 11
1.4 Particle Size Analysis . . . . . . . . . . . . . . . 12
1.5 Density Properties for Granular Soils . . . . . . . . 14
1.6 Plasticity Properties of Soils . . . . . . . . . . . . 18
1.7 Soil Compaction . . . . . . . . . . . . . . . . . . 23
1.8 Engineering Classification of Soils . . . . . . . . . . 26
Exercises . . . . . . . . . . . . . . . . . . . . . . . 28
Bibliography . . . . . . . . . . . . . . . . . . . . . . 29

**2. Permeability of Soil and Seepage Force** **31**

2.1 Introduction . . . . . . . . . . . . . . . . . . . . 32
2.2 Capillary Phenomena . . . . . . . . . . . . . . . . 33

2.3 Groundwater Movement and Darcy's Law . . . . . . 34
2.4 Determination of Permeability Coefficient . . . . . . 38
2.5 Flow Nets . . . . . . . . . . . . . . . . . . . . . . 46
2.6 Seepage Force . . . . . . . . . . . . . . . . . . . . 47
2.7 Critical Hydraulic Gradient . . . . . . . . . . . . . 50
Exercises . . . . . . . . . . . . . . . . . . . . . . . . 52
Bibliography . . . . . . . . . . . . . . . . . . . . . . 53

**3. Stress Distribution in Soil** **55**

3.1 Introduction . . . . . . . . . . . . . . . . . . . . . 56
3.2 Stresses Due to Self-Weight . . . . . . . . . . . . . 60
3.3 Effective Stress Principle . . . . . . . . . . . . . . 62
3.4 Contact Pressure between the Foundation and the Ground . . . . . . . . . . . . . . . . . . 65
3.5 Additional Stress in Ground Base . . . . . . . . . . 73
3.6 Additional Stress in Plane Problem . . . . . . . . . 87
Exercises . . . . . . . . . . . . . . . . . . . . . . . . 96
Bibliography . . . . . . . . . . . . . . . . . . . . . . 100

**4. Compression and Consolidation of Soils** **101**

4.1 Introduction . . . . . . . . . . . . . . . . . . . . . 102
4.2 Compressibility Characteristics . . . . . . . . . . . 102
4.3 Calculation of Settlement of Foundation . . . . . . . 108
4.4 One-Dimensional Consolidation Theory . . . . . . . 112
Exercises . . . . . . . . . . . . . . . . . . . . . . . . 119
Bibliography . . . . . . . . . . . . . . . . . . . . . . 120

**5. Shear Strength** **123**

5.1 Shear Resistance . . . . . . . . . . . . . . . . . . . 124
5.2 Mohr–Coulomb Failure Criterion . . . . . . . . . . 126
5.3 Shear Strength Tests . . . . . . . . . . . . . . . . . 130
5.4 Effective Stress Paths . . . . . . . . . . . . . . . . 130
5.5 Characteristics of Shear Strength of Cohesionless Soil . . . . . . . . . . . . . . . . . . . 132
5.6 Characteristics of Shear Strength of Cohesive Soil . 135
Exercises . . . . . . . . . . . . . . . . . . . . . . . . 141
Bibliography . . . . . . . . . . . . . . . . . . . . . . 141

**6. Bearing Capacity** **143**

6.1 Introduction . . . . . . . . . . . . . . . . . . . . . . . . 144
6.2 Critical Edge Pressure . . . . . . . . . . . . . . . . . . 146
6.3 Prandtl's Bearing Capacity Theory . . . . . . . . . . 149
6.4 Modification of Prandtl's Bearing Capacity Theory . 150
6.5 Terzaghi's Bearing Capacity Theory . . . . . . . . . 152
Exercises . . . . . . . . . . . . . . . . . . . . . . . . . . . 158
Bibliography . . . . . . . . . . . . . . . . . . . . . . . . . 158

**7. Slope Stability Analysis** **161**

7.1 Introduction . . . . . . . . . . . . . . . . . . . . . . . . 162
7.2 Factors and Engineering Measures that Influence Soil Slope Stability . . . . . . . . . . . . . . . . . . . 163
7.3 Stability Analysis of Cohesionless Soil Slope . . . . . 166
7.4 Stability Analysis of a Cohesive Soil Slope . . . . . . 168
7.5 Stability Number Method . . . . . . . . . . . . . . . 181
Exercises . . . . . . . . . . . . . . . . . . . . . . . . . . . 183
Bibliography . . . . . . . . . . . . . . . . . . . . . . . . . 185

**8. Earth Pressure and Retaining Walls** **187**

8.1 Introduction . . . . . . . . . . . . . . . . . . . . . . . . 188
8.2 Earth Pressure on the Retaining Wall . . . . . . . . . 188
8.3 Calculation of the At-Rest Earth Pressure . . . . . . 191
8.4 Rankine's Earth Pressure Theory . . . . . . . . . . . 193
8.5 Coulomb's Earth Pressure Theory . . . . . . . . . . . 200
8.6 Design of Earth-Retaining Structures . . . . . . . . . 215
Exercises . . . . . . . . . . . . . . . . . . . . . . . . . . . 220
Bibliography . . . . . . . . . . . . . . . . . . . . . . . . . 221

*Appendix A: English–Chinese–Japanese Translation of Frequently used Terminologies* 223

*Appendix B: Answers to the Exercises* 235

*Index* 239

# List of Symbols

$a$ compression coefficient
$b$ width of foundation
$c$ cohesion
$c'$ effective cohesion
$C_c$ compression index of soil
$C_e$ rebound index
$C_u$ coefficient of uniformity
$C_v$ coefficient of consolidation
$d$ depth of foundation
$d_s$ specific gravity
$D$ average diameter of the soil particle
$D_r$ relative density
$D_{10}$ effective size
$D_{30}$ continuous size
$D_{50}$ average size
$D_{60}$ constrained size
$e$ void ratio
$E$ Young's modulus
$E_0$ at-rest earth pressure
$E_a$ active earth pressure
$E_p$ passive earth pressure
$E_s$ oedometric modulus
$f_{ak}$ characteristic value of subgrade bearing capacity
$g$ acceleration of gravity

| | |
|---|---|
| $G$ | shear modulus |
| $G_{\mathrm{d}}$ | seepage force |
| $h$ | water head |
| $H$ | height of slope |
| $H_{\mathrm{cr}}$ | critical safety height of slope |
| $i$ | hydraulic gradient |
| $i_{\mathrm{cr}}$ | critical hydraulic gradient |
| $I_{\mathrm{c}}$ | consistency index |
| $I_{\mathrm{L}}$ | liquidity index |
| $I_{\mathrm{p}}$ | plasticity index |
| $k$ | coefficient of permeability |
| $K_0$ | at-rest earth pressure coefficient |
| $K_{\mathrm{a}}$ | coefficient of active earth pressure |
| $K_{\mathrm{p}}$ | coefficient of passive earth pressure |
| $K_{\mathrm{s}}$ | safety factor or factor of safety of anti-sliding stability |
| $K_{\mathrm{t}}$ | factor of safety of anti-overturning stability |
| $m$ | mass |
| $M_{\mathrm{r}}$ | anti-sliding moment |
| $M_{\mathrm{s}}$ | sliding moment |
| $n$ | porosity |
| $N_{\mathrm{s}}$ | stability number |
| $N_{\gamma}, N_{\mathrm{q}}, N_{\mathrm{c}}$ | bearing capacity factor |
| OCR | overconsolidation ratio |
| $p$ | contact pressure |
| $p_{\mathrm{cr}}$ | critical edge pressure |
| $p_{\mathrm{u}}$ | ultimate bearing capacity |
| $q$ | deviator stress |
| $q_{\mathrm{u}}$ | unconfined compressive strength |
| $Q$ | rate of seepage |
| $R_n$ | Reynolds number |
| $s$ | settlement |
| $S_{\mathrm{r}}$ | degree of saturation |
| $T$ | sliding force |
| $u$ | pore water pressure |
| $U_{\mathrm{t}}$ | degree of consolidation |
| $v$ | discharge velocity |
| $V$ | volume |
| $w$ | water content |

$w_L$ liquid limit
$w_p$ plastic limit
$w_s$ shrinkage limit
$\alpha$ compression coefficient
$\beta$ slope angle
$\gamma_d$ dry unit weight
$\gamma_{sat}$ saturated unit weight
$\gamma_0$ average weight above the base
$\delta$ friction angle between retaining wall and soil
$\nu$ Poisson's ratio
$\rho$ bulk density
$\rho_d$ dry density
$\rho_{sat}$ saturated density
$\rho'$ buoyant density
$\sigma'$ effective stress
$\sigma_{cz}$ geostatic stress
$\sigma_z$ additional stress
$\sigma$ stress
$\sigma_1$ maximum principal stress
$\sigma_2$ intermediate principal stress
$\sigma_3$ minimum principal stress
$\tau_f$ shear strength
$\varphi$ internal friction angle
$\varphi'$ effective internal friction angle
$\gamma$ unit weight

# Introduction

## Importance of This Course

Soil mechanics is a basic and mandatory course for undergraduate students in civil engineering. Students who take up this course can master the basic theory and skills of soil mechanics related to civil engineering. Soil mechanics is closely related to the construction branch in civil engineering. Geological conditions of the site and mechanical behavior of the soil are vital points to keep in mind before constructing dams, embankments, tunnels, canals and waterways, foundations for bridges, roads, buildings, and solid waste disposal systems.

The geological structure of China is diverse. There are various types of soils, the properties of which also vary significantly. Furthermore, some special soils or regional soils (such as soft soil, collapsible loess, expansive soil, red clay, and permafrost) exhibit behaviors different from general soils. Therefore, it is necessary to study the characteristics of soils in order to take appropriate engineering measures. There are many geotechnical problems caused by natural factors or human intervention, which involve various engineering activities, such as civil engineering, mining engineering, and underground engineering. The design and construction of engineering structures depend on the geological environment and mechanical behavior of soils. Therefore, it is necessary to understand the characteristics (such as physical, chemical and mechanical) of soils.

## Characteristics and Engineering Background of the Course

As an important basic application-oriented course and as a practical science course in engineering, soil mechanics is useful in studying the characteristics and engineering behavior of soils. It is useful in analyzing and solving the engineering problems encountered in the design and construction of the foundation and problems associated with geomaterials. This course is an important part of the civil engineering disciplines.

Soil mechanics is a practical engineering discipline that takes soils as the research object and is considered as a branch of engineering mechanics. Weathered rocks disintegrate, metamorphose, get transported to a new environment by various natural forces, and get accumulated or deposited there, thereby forming soils. Together with a knowledge of mechanics and engineering geology, soil mechanics can be useful in studying the stress, strain, strength and stability of soils in relation to construction engineering under the action of external factors (such as load, water, temperature) using the principles of mechanics and geotechnical testing techniques. Therefore, soil mechanics is a very practical engineering science.

The practice of utilizing soils can be traced back to ancient times. Our ancestors used soil as construction material to build burial sites, flood protection bunds, and shelters. From the Neolithic sites found in Banpo village, Xi'an of China, it is seen that, at that time, people had used soil pedestal and stone foundation for constructing simple houses taking into account foundation stability problems as well. During the rule of the Qin dynasty compaction method was used to build roads, and during the rule of the Sui and Tang dynasties, wood-pile and lime soil foundation, respectively, were used to build towers. During the European industrial revolution, with developments in construction industry, advancements in railway and highway sectors, and the progress in science, the theory of soil mechanics was established. The first scientific study of soil mechanics was undertaken by the French physicist Charles-Augustin de Coulomb, who proposed an equation for calculating the shear strength of sand and developed a theory for earth pressure in the year 1773. Coulomb's theory and the theory of earth pressure published by the Scottish engineer William Rankine in 1857 are used even now to quantify earth pressures.

In 1869, Карлович published the world's first book on foundation. According to the theory of elasticity, J. Boussinesq obtained the analytic solution of three-dimensional stress distribution in a foundation under the action of a concentrated load in 1885. In 1900, C. O. Mohr proposed the soil strength theory. By the early 20th century, people had gathered a lot of experience and accumulated a large amount of data in engineering practice and started theoretical discussions on the strength, deformation, and permeability properties of soil. Thus, soil mechanics gradually formed as an independent discipline. In 1920, L. Prandtl proposed the bearing capacity theory. In this period, great developments occurred in the theory of slope stability. W. Fellenius improved the analysis method for circular sliding of slopes, known as the Fellenius method of slices or the Swedish method of slices. Based on the practical and theoretical investigations done by professionals and researchers in civil engineering for more than a century, in 1925, K. Terzaghi summarized and published the first book on *Soil Mechanics* in United States. In 1929, he published the book *Engineering Geology* in collaboration with other writers. Since then, soil mechanics, engineering geology, and foundation engineering have continuously developed as independent sciences. Since the first session in 1936 in USA, a total of 18 International Conferences on Soil Mechanics and Geotechnical Engineering have been conducted till 2013. Scientists from around the world exchanged research experience on this subject during the conference. With the developments and the advancements in science, more and more theories and techniques have been developed in soil mechanics. Application and improvement of basic characteristics of soil, effective stress principle, consolidation theory, deformation theory, stability soil mass, soil dynamics, soil rheology, etc., in soil mechanics were the main topics in this field. In 1954, В. В. Соколовский published a book titled *Loose Media Statics*. A. W. Skempton, A. W. Bishop, and N. Janbu had made contributions to the effective stress principle and the theory of slope stability. Chinese scholars Wenxi Huang, Zongji Chen, Jiahuan Qian, and Zhujiang Shen had made significant contributions to constitutive relations of soil, clay microstructure and soil rheology, geotechnical earthquake engineering and soil rheology, and constitutive relations of soft soil, respectively.

The composition and engineering geological conditions of the supporting soil for foundations are complex and different from each other. The requirements of engineering geology for the supporting soil are different for different building structures. Therefore, the engineering problems of soils are varied in nature. The composition, thickness, physical and mechanical properties, bearing capacity of soils, etc., are the basic conditions for assessing the stability of the soil supporting the foundation. Therefore, degradation of the supporting soil is a frequently encountered problem in construction engineering. A well-known example was the instability of the supporting soil for a grain elevator constructed in Transcona, Canada. Another example is the non-uniform settlement of the foundation for the Leaning Tower of Pisa in Italy.

The Transcona grain elevator completed in September 1913 was 59.44 m long, 31 m high, and 23.47 m wide. Within 24 h after loading the grain elevator at a rate of about 1 m of grain height per day, the bin house began to tilt and settle. Fortunately, the structural damage was minimal and the bin house was later restored. No borings were done to identify the soil type and to obtain information on their strength.

The Tower of Pisa is located in the city of Pisa, Italy. The city is located on the banks of the Arno River, northwest to Rome. The tower is 54 m in height and 142,000 kN in weight. The Pisa Tower was built in several stages from 1173 to 1370. During this period, the construction stopped twice due to inclination. Prior to restoration work performed between 1990 and 2001, the tower leaned at an angle of 5.5°. The tower now leans at about 3.99°. This means that the top of the tower is horizontally displaced to a distance of 3.9 m from where it would originally be if the structure were perfectly vertical. The tower started tilting during the construction stage itself because the south side of the soil mass supporting the foundation was too soft to properly support the structure's weight. Plastic deformation of the foundation, creep, falling water tables, etc., have been the reasons for the accelerated rate of inclination of the tower. Circular excavations were made for unloading at the opposite side, and grouting was done to reinforce the soil surrounding the foundation. The body of the tower was also reinforced to prevent its collapse.

To ensure the stability and serviceability of a building, the bearing capacity of the supporting soil must meet two basic conditions: strength and deformation. The soil should have sufficient strength to ensure the stability of the ground under loading conditions.

On the contrary, the deformation of the ground should not exceed the permissible value for the building. Therefore, an ideal ground for construction generally should have higher strength and lower compressibility.

## Fundamental Characteristics and Requirements

The fundamental characteristics and requirements of each part are as follows:

(1) **The basic properties and the permeability of soil:** Students are required to understand the concept of a three-phase composition of the soil, parameters for physical properties of soil and their relationship, soil permeability, the theoretical background of Darcy's law and methods for measuring the coefficient of permeability, concepts and calculation of hydrodynamic pressure and critical hydraulic gradient, the main types of seepage failure and its prevention measures, and the engineering classification of soil.

(2) **Stress distribution in soil:** The stress in soils will be redistributed during the construction, which causes deformation of the ground. If the induced stress is too large and exceeds the ultimate bearing capacity of soils, it will cause failure of the ground. Therefore, understanding the calculation of stress and deformation is the premise for ensuring the serviceability and safety of buildings. Students are required to master the basic concepts of effective stress, pore water pressure, gravity stress and additional stress, the theory of effective stress under both hydrostatic and seepage conditions, the calculation methods for gravity stress, effective stress, foundation pressure, and additional stresses.

(3) **Compression and consolidation of soil:** Stress and deformation occur as a result of the load of buildings, which eventually lead to subsidence and non-uniform settlement of the foundation. If the settlement of the foundation exceeds a certain limit, it will cause deformation, cracking, tilting, or even collapse of buildings. Therefore, the calculation of the settlement is an important issue related to safety and stability of a building. Thus, students are required to understand the compression characteristics of

soils and the consolidation state, the compressive index of soil and its determination method, the layerwise summation method for calculating the foundation settlement, and the foundation settlement calculation methods for normally consolidated and overconsolidated soils, including calculation methods of one-way seepage consolidation settlement of the foundation. These are the fundamental criteria that are required for settlement calculation and the foundation design of actual engineering structures.

(4) **Shear strength of soil and bearing capacity:** Shear strength is one of the important mechanical properties of the soil, which is closely related to the stability and serviceability of buildings. Determination of bearing capacity is the basic requirement before designing a foundation. Students are required to understand the Mohr–Coulomb theory and the limit equilibrium condition of the soil, the shear strength indexes and test methods for determining these indexes, shear strength properties of soil, failure characteristics of soils, and the calculation methods for critical edge loads and ultimate loads.

(5) **Slope stability and earth pressure:** Earth pressure and slope stability are the problems frequently encountered in construction engineering, which must also be analyzed for the engineering design and construction. Students are required to understand the basic concepts and calculation methods for earth pressure, Rankine's earth pressure theory and Coulomb's earth pressure theories, the types of retaining structures, factors affecting the slope stability, the slope stability analysis for non-cohesive and cohesive soils, the Swedish method of slices, and Bishop's method of circular sliding surface.

# Chapter 1

# Basic Characteristics and Engineering Classification of Soils

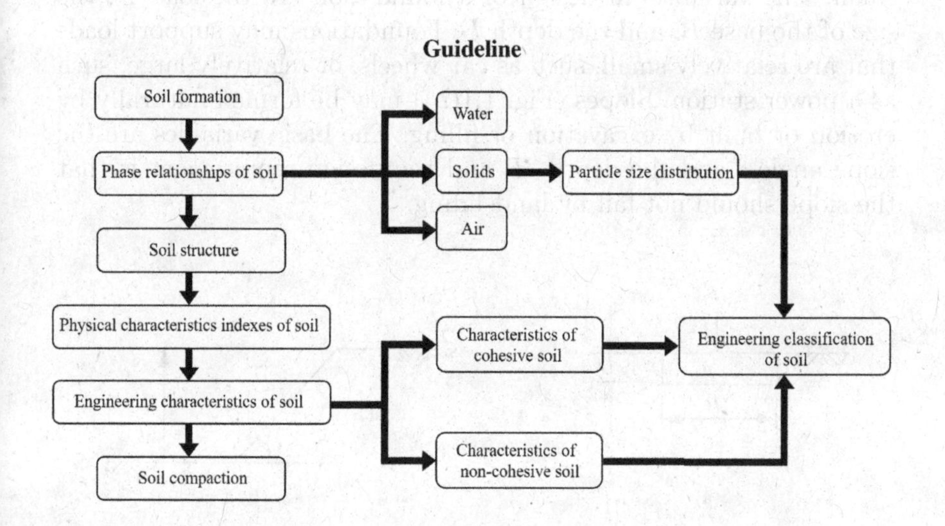

## 1.1 Introduction

Soil mechanics may be defined as the study of the engineering behaviors of soils that are related to the design of civil engineering structures made from or in the earth. Nearly all of the civil engineering structures, such as buildings, bridges, highways, tunnels, earth retaining walls, embankments, basements, sub-surface waste repositories, towers, canals, and dams, are constructed in or on the surface of the earth. To perform satisfactorily, each structure must have a proper foundation.

The four basic types of geotechnical structures are illustrated in Fig. 1.1, and most of the other cases are variations or combinations of these. Foundations (Fig. 1.1(a)) transmit loads to the ground and the basic criterion for design is that the settlements should be relatively small. The variables in design of a foundation are the load $F$, the size of the base $B$, and the depth $D$. Foundations may support loads that are relatively small, such as car wheels, or relatively large, such as a power station. Slopes (Fig. 1.1(b)) may be formed naturally by erosion or built by excavation or filling. The basic variables are the slope angle $i$ and the depth $H$, and the design requirement is that the slope should not fail by landsliding.

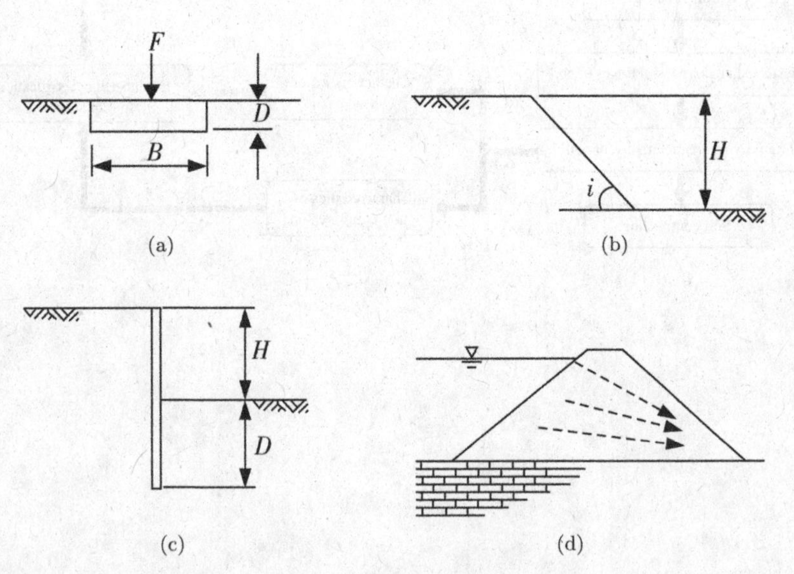


Fig. 1.1. Geotechnical structures. (a) Foundation, (b) slope stability, (c) retaining wall, and (d) earth–fill dam.

Stand slopes that are deep and steep can be supported by a retaining wall (Fig. 1.1(c)). The basic variables are the height of the wall $H$ and its depth of burial $D$, together with the strength and stiffness of the wall and the forces in any anchors or props. The basic requirements for the design are complex and involve overall stability, restriction of ground movements, and the bending and shearing resistance of the wall. In any structure where there are different levels of water, such as in a dam (Fig. 1.1(d)) or around a pumped well, there will be seepage of water. The seepage causes leakage through a dam and governs the yield of a well and it also governs the variation of pressure in the groundwater.

The structures are stable, taking into consideration that the safety of design section, the soils that have problems such as ploughing, flow of mineral ore, and grain from storage soil can be solved by the theory of soil mechanics. Other problems in geotechnical engineering include movement of contaminations from the place of depositing the waste and techniques for ground improvement.

## 1.2 Soil Composition and Phase Relationships

### 1.2.1 *Soil composition*

Soil is a particulate material, which means that a soil mass consists of an accumulation of individual particles that are bonded together by mechanical or attractive means, though not as strongly as for rock. To the civil engineer, soil is an uncemented or weakly cemented accumulation of mineral particles formed by the weathering of rocks, the void space between the particles containing water and/or air.

In nature, the void of every soil is partly or completely filled with water. Soils can be of either two phases or three phases composition, as illustrated in Fig. 1.2. A partially saturated soil has three phases, being composed of solid soil particles, pore water, and pore air. In a completely dry soil, there are two phases, namely the solid soil particles and pore air. A fully saturated soil also has two phases, being composed of solid soil particles and pore water. Below the water table, the soil is assumed to be fully saturated, although it is likely that, due to the presence of small volumes of entrapped air, the degree of saturation will be marginally below 100%.

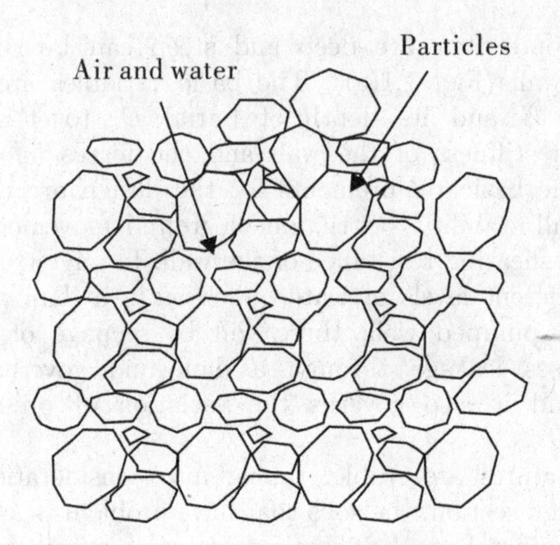


Fig. 1.2. Three-phase composition.

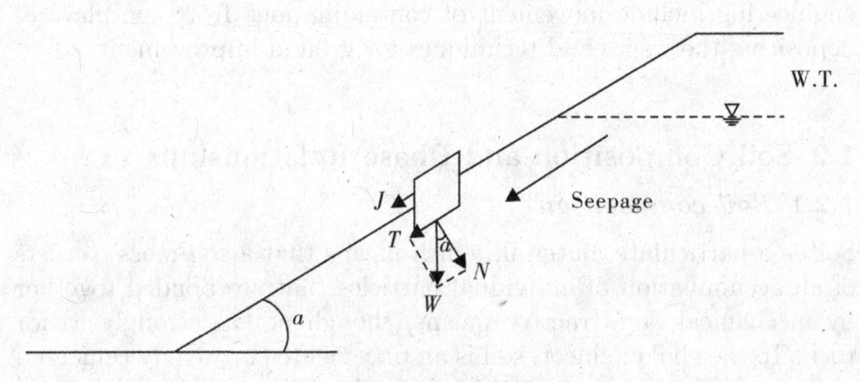


Fig. 1.3. Seepage of slope.

The water filling the void space may be in a state of rest or in a state of flow. Below the water table, the pore water may be static, the hydrostatic pressure depending on the depth below the water table, or may be seeping through the soil under hydraulic gradient, as illustrated in Fig. 1.3. If the water is in a state of rest, the methods for solving stability and deformation problems are essentially identical with those for solving similar problems in the mechanics of solids in general. On the other hand, if the water percolates through the voids of the soil, the problems cannot be solved without previously determining the state of stress in the water contained in the voids of the soil. In this case, we are obliged to combine the mechanics of solids with applied hydraulics.

Significant engineering properties of a soil deposit, such as strength and compressibility, are directly related to or at least affected by basic factors such as volume and weight of a bulk soil consisting of solid particles or water or air. Information such as soil density (weight per unit volume), water content, void ratio, and degree of saturation are used in calculations to determine the stability of earth slopes, foundation settlement, and the bearing capacity of foundations. Therefore, we can determine the suitability of foundation types and constructions. For this reason, an understanding of the terminology and definitions related to soil composition is the key to studying soil mechanics.

### 1.2.2 *Phase relationships*

The components of a soil can be represented by a phase diagram as shown in Fig. 1.4. Bulk soil as it exists in nature is a more or less random accumulation of soil particles, water, and air space, as shown in Fig. 1.4(a). For purposes of study and analysis, it is convenient to represent this soil mass by a phase or block diagram, which consists of three parts: solid particles, water or other liquid, and air or other gas, as shown in Fig. 1.4(b).

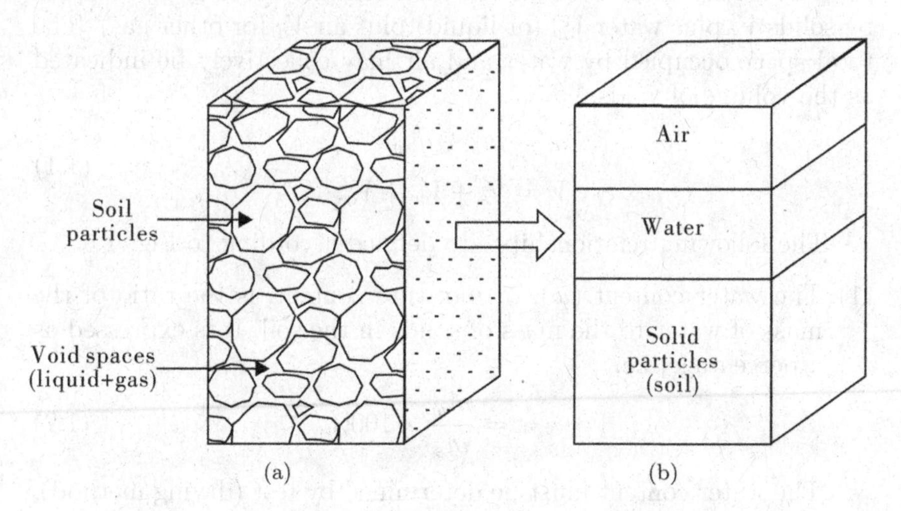


Fig. 1.4. Phase diagrams.

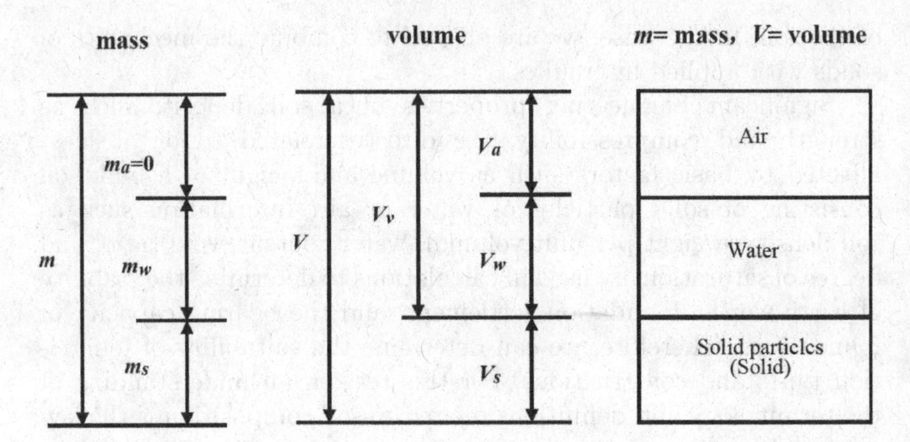


Fig. 1.5. Phase relationships.

The phase relationships as shown in Fig. 1.5 illustrate the mass and volume of components of soil system. The relationships are summarized in Fig. 1.5. The total mass $m$ of the soil volume is taken as the sum of the mass of solids $m_s$ and water $m_w$. The mass of air (in the voids) measured in air (the earth's atmosphere) is zero. The air or other gases may have a measurable weight, but it would normally be very small compared to the total mass of soil plus water and therefore can be neglected without causing serious error.

The total volume $V$ of the soil bulk includes the volume occupied by solids $V_s$ plus water $V_w$ (or liquid) plus air $V_a$ (or other gas). The total space occupied by water and air may collectively be indicated as the volume of voids.

$$\begin{aligned} m &= m_s + m_w \\ V &= V_s + V_w + V_a. \end{aligned} \tag{1.1}$$

The following relationships are defined according to Fig. 1.5.

(1) The water content ($w$), or moisture content, is the ratio of the mass of water to the mass of solids in the soil. It is expressed as a percentage, i.e.

$$w = \frac{m_w}{m_s} \times 100\%. \tag{1.2}$$

The water content must be determined by test (drying method). It is determined by weighing a sample of the soil and then drying the sample in an oven at a temperature of 105–110°C and reweighing.

(2) The bulk density ($\rho$) of a soil is the ratio of the total mass to the total volume, i.e.

$$\rho = \frac{m}{V}. \tag{1.3}$$

Convenient units for density are kg/m$^3$ or g/cm$^3$. The density of water $\rho_w = 1000\,\text{kg/m}^3$ (or $1.00\,\text{g/cm}^3$).
For a completely dry soil ($S_r = 0$),

$$\rho_d = \frac{m_s}{V}. \tag{1.4}$$

For a fully saturated soil ($S_r = 1$),

$$\rho_{\text{sat}} = \frac{(m_s + V_v \rho_w)}{V}. \tag{1.5}$$

The buoyant density ($\rho'$) is given by

$$\rho' = \frac{(m_s - V_s \rho_w)}{V}. \tag{1.6}$$

The unit weight ($\gamma$) of a soil is the ratio of the total weight (a force) to the total volume, i.e.

$$\gamma = \frac{W}{V} = \frac{mg}{V} = \rho g. \tag{1.7}$$

Convenient units are kN/m$^3$.

(3) The specific gravity of the soil particles ($d_s$) is given by

$$d_s = \frac{m_s}{V_s \rho_w} = \frac{\rho_s}{\rho_w}, \tag{1.8}$$

where $\rho_s$ is the particle density. If the units of $\rho_s$ are g/cm$^3$, then $\rho_s$ and $d_s$ are numerically equal.

(4) The void ratio ($e$) is the ratio of the volume of voids to the volume of solids, i.e.

$$e = \frac{V_v}{V_s}. \tag{1.9}$$

The porosity ($n$) is the ratio of the volume of voids to the total volume of the soil, i.e.

$$n = \frac{V_v}{V} \times 100\%. \tag{1.10}$$

The $e$ and the $n$ are inter-related as follows:

$$e = \frac{n}{1-n} \tag{1.11}$$

$$n = \frac{e}{1+e} \times 100\%. \tag{1.12}$$

(5) The degree of saturation ($S_r$) is the ratio of the volume of water to the total volume of void space. It is also expressed as a percentage, i.e.

$$S_r = \frac{V_w}{V_v} \times 100\%. \tag{1.13}$$

The degree of saturation can range between the limits of zero for a completely dry soil and 1 (or 100%) for a fully saturated soil.

From the definition of void ratio, if the volume of solids is 1 unit, then the volume of voids is $e$ units. The mass of solids is then $d_s$ and, from the definition of water content, the mass of water is $d_s w \rho_w$. The volume of water is thus $d_s w$. These volumes and masses are represented in Fig. 1.6. The following relationships can be obtained:

$$e = \frac{d_s(1+w)\rho_w}{\rho} - 1 \tag{1.14}$$

$$S_r = \frac{d_s w}{e} \tag{1.15}$$

$$\rho = \frac{d_s(1+w)\rho_w}{1+e} \tag{1.16}$$

$$\rho_{\text{sat}} = \frac{d_s + e}{1+e}\rho_w \tag{1.17}$$

$$\rho_d = \frac{d_s}{1+e}\rho_w. \tag{1.18}$$

Table 1.1 shows the formulas of physical indexes.

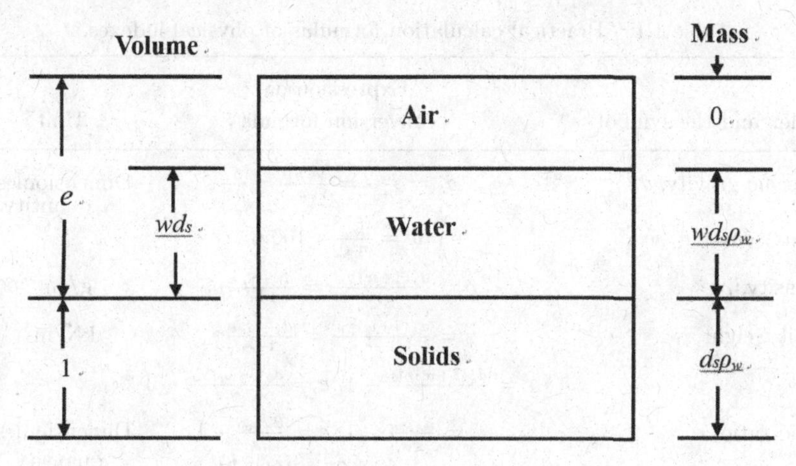


Fig. 1.6. Relationship of volumes and masses.

**Example 1.1.** In its natural condition, a soil sample has a mass of 2290 g and a volume of $1.15 \times 10^{-3}$ m$^3$. After being completely dried in an oven, the mass of the sample is 2035 g. The value of $d_s$ for the soil is 2.68. Determine the water content, bulk density, unit weight, void ratio, porosity, and degree of saturation.

**Solution:** According to Eqs. (1.2), (1.3), and (1.7), the water content, bulk density, and unit weight can be calculated directly by definition.

$$\text{Water content,} \quad w = \frac{m_w}{m_s} = \frac{2290 - 2035}{2035} \times 100\% = 12.5\%$$

$$\text{Bulk density,} \quad \rho = \frac{m}{V} = \frac{2.29}{1.15 \times 10^{-3}} = 1990 \text{ kg/m}^3$$

$$\text{Unit weight,} \quad \gamma = \frac{mg}{V} = 1990 \times 9.8 = 19.5 \text{ kN/m}^3.$$

The void ratio, porosity, and degree of saturation need to use phase relationships, by Eqs. (1.14), (1.12), and (1.15):

$$\text{Void ratio,} \ e = \frac{d_s(1+w)\rho_w}{\rho} - 1 = \frac{2.68 \times 1.125 \times 1000}{1990} - 1 = 0.52$$

Table 1.1. Practical calculation formulas of physical indexes.

| Index and the symbol | Expression or conversion formula | Unit |
|---|---|---|
| Specific gravity, $d_s$ | $d_s = \frac{m_s}{m_w} = \frac{V_s\rho_s}{V_s\rho_w} = \frac{\rho_s}{\rho_w}$ | Dimensionless quantity |
| Water content, $w$ | $w = \frac{m_w}{m_s} \times 100\%$ | % |
| Density, $\rho$ | $\rho = \frac{d_s(1+w)\rho_w}{1+e} = \frac{d_s+S_r e}{1+e}\rho_w$ | g/cm$^3$ |
| Unit weight, $\gamma$ | $\gamma = \frac{d_s(1+w)\gamma_w}{1+e} = \frac{d_s+S_r e}{1+e}\gamma_w$ | kN/m$^3$ |
| Void ratio, $e$ | $e = \frac{d_s(1+w)\rho_w}{\rho} - 1$, $e = \frac{d_s(1+w)\gamma_w}{\gamma} - 1$<br>$e = \frac{d_s\rho_w}{\rho_d} - 1$, $e = \frac{d_s\gamma_w}{\gamma_d} - 1$<br>$e = d_s w$ (当$S_r = 100\%$时) | Dimensionless quantity |
| Porosity, $n$ | $n = \frac{e}{1+e} \times 100\%$ | % |
| Degree of saturation, $S_r$ | $S_r = \frac{d_s w}{e}$ | % |
| Dry density, $\rho_d$ | $\rho_d = \frac{d_s}{1+e}\rho_w$, $\rho_d = \frac{\rho}{1+w}$ | g/cm$^3$ |
| Dry unit weight, $\gamma_d$ | $\gamma_d = \frac{d_s}{1+e}\gamma_w$, $\gamma_d = \frac{\gamma}{1+w}$ | kN/m$^3$ |
| Saturated density, $\rho_{sat}$ | $\rho_{sat} = \frac{d_s+e}{1+e}\rho_w$ | g/cm$^3$ |
| Saturated unit weight, $\gamma_{sat}$ | $\gamma_{sat} = \frac{d_s+e}{1+e}\gamma_w$ | kN/m$^3$ |
| Buoyant density, $\rho'$ | $\rho' = \frac{d_s-1}{1+e}\rho_w$, $\rho' = \rho_{sat} - \rho_w$<br>$\rho' = (d_s-1)(1-n)\rho_w$ | g/cm$^3$ |
| Buoyant unit weight, $\gamma'$ | $\gamma' = \frac{d_s-1}{1+e}\gamma_w$, $\gamma' = \gamma_{sat} - \gamma_w$<br>$\gamma' = (d_s-1)(1-n)\gamma_w$ | kN/m$^3$ |

$$\text{Porosity, } n = \frac{e}{1+e} = \frac{0.52}{1.52} \times 100\% = 34\%$$

$$\text{Degrees of saturation, } S_r = \frac{d_s w}{e} = \frac{2.68 \times 12.5\%}{0.52} = 64.5\%.$$

**Example 1.2.** A dry soil is measured with bulk density $\rho = 1.69 \times 10^3$ kg/m$^3$, specific gravity $d_s = 2.70$. After a rain, the total volume of the soil doesn't change and the degree of saturation $S_r$ becomes 40%.

Calculate bulk density $\rho$ and water content $w$ of the soil after the rain.

**Solution:** For a dry soil, the dry density $\rho_d$ is numerically equal to the bulk density $\rho$, that is $\rho_d = 1.69 \times 10^3\,\text{kg/m}^3$, then from Eq. (1.18),

$$\text{Void ratio, } e = \frac{d_s \rho_w}{\rho_d} - 1 = \frac{2.70 \times 1000}{1690} - 1 = 0.6.$$

Therefore,

$$\text{Water content after the rain, } w = \frac{S_r e}{d_s} = \frac{40\% \times 0.6}{2.7} = 8.89\%$$

Bulk density after the rain,

$$\rho = \frac{d_s(1+w)\rho_w}{1+e} = \frac{2.7 \times (1 + 8.89\%) \times 1000}{1 + 0.6}$$
$$= 1.84 \times 10^3\,\text{kg/m}^3.$$

## 1.3 Soil Fabric

In the micro scale, soil fabric is formed by the certain structure of soil particles. When it comes to soil fabric, it refers to particle size, shape, arrangement and especially the mutual connecting condition, which is synthetical of soil grain characteristics. Figure 1.7 shows the three kinds of basic types of soil fabrics.

Crushed stone, gravel, sand, and some other coarse grain soil all belong to single grain fabric, which is formed in sedimentation. According to its arrangement situation, it can be divided into two types — the tight and the loose, as shown in Fig. 1.7(a). The greater the range of particle sizes present, the tighter the structure can be. The grain of sand or gravel is big and relative surface area and surface energy are quite small compared to its gravity. There is only strong bound water on particle surface and almost no inter-particle force among particles. Thus, this kind of soil is also called cohesionless soil.

Silt belongs to honeycomb fabric, as shown in Fig. 1.7(b). In the sedimentation process, where the fine grained soils contact the soils which have already sunk, it will stay at the points of contact and

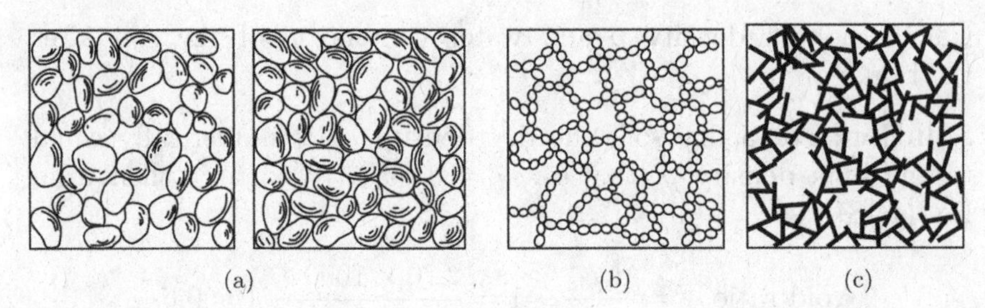


Fig. 1.7. Soil fabric. (a) Single grain fabric, (b) honeycomb fabric, and (c) flocculent fabric.

not sink, for which the interparticle force is bigger than the gravity. Thus, the honeycomb fabric which has big voids comes into being, as can be seen in Fig. 1.7(b). This kind of soil is quite unsteady and will produce great deformation under water immersion or external force.

Clay particles belong to flocculent fabric, as shown in Fig. 1.7(c). Clay ($d < 0.005$ mm), colloidal particles, and clay mineral particles ($d < 0.002$ mm) are very fine, presenting thin-bedded state, which makes this kind of soil have a large relative surface area and be always suspended in water. If the interparticle reaction appears to be an attractive force, they easily combine with each other and gradually form a little link aggregation of soil grains and make the mass bigger and then sink. When a link touches the other, they attract each other, and gradually a big link can be formed. The soil grain of this kind of fabric arranges randomly and has big pore space, as a result, it will show low strength and high compressibility, meanwhile, it is sensitive to disturbance. But the linkage strength among soil grains is becoming bigger and bigger for compaction and cementation, which is the main source of cohesive force in clay.

## 1.4 Particle Size Analysis

The particle size analysis of a soil sample involves determining the percentage by mass of particles within the different size ranges. Particle sizes in soils can vary from less than 0.001 mm to over 100 mm. By the Chinese, British, and Japanese Standards, the size ranges detailed in Fig. 1.8 are specified. The same terms are also used to describe particular types of soil. The particle size distribution of a coarse-grained soil can be determined by the method of sieving. The

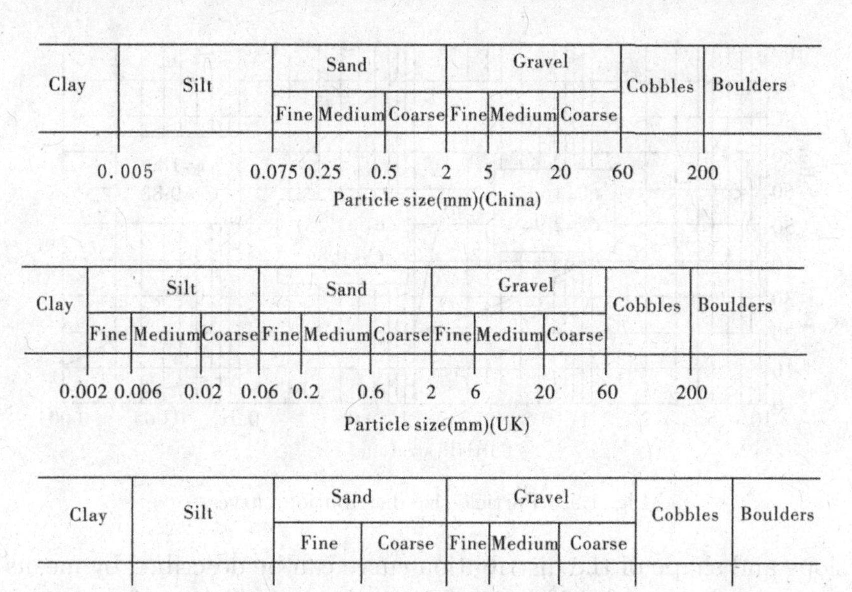


Fig. 1.8. Particle size ranges.

particle size distribution of a fine-grained soil or the fine-grained fraction of a coarse-grained soil can be determined by the method of sedimentation.

### 1.4.1 *Method of sieving*

The soil sample is passed through a series of standard test sieves having successively smaller mesh sizes. The mass of soil retained in each sieve is determined and the cumulative percentage by mass passing each sieve is calculated. The particle size distribution of a soil is presented as a curve on a semilogarithmic plot. The ordinate is the percentage by mass of particles smaller than the size given by the abscissa. Examples of particle size distribution curves appear in Fig. 1.9. The particle size corresponding to any specified value on the "percentage smaller" scale can be read from the particle size distribution curve.

The size in which 10% of the particles are smaller than that size is denoted by $D_{10}$. Other sizes such as $D_{30}$ and $D_{60}$ can be defined in a similar way. The size $D_{10}$ is defined as the effective size. The general

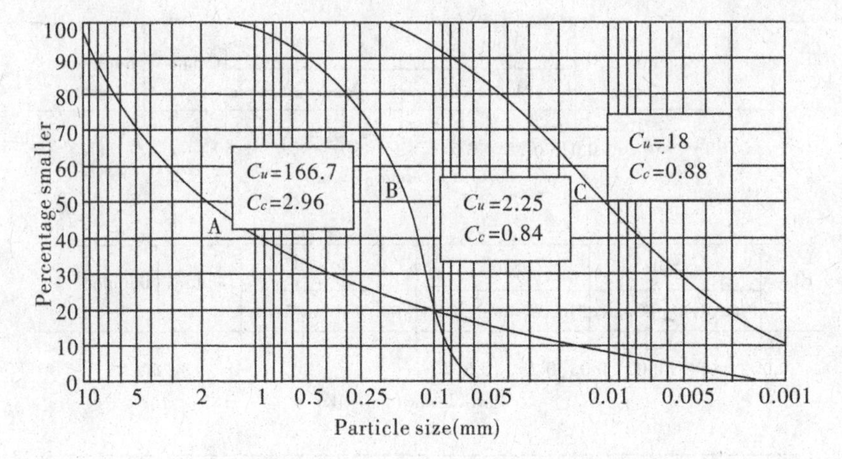


Fig. 1.9. Particle size distribution curves.

slope and shape of the distribution curve can be described by means of the coefficient of uniformity ($C_u$) and the coefficient of curvature ($C_c$), defined as follows:

$$C_u = \frac{D_{60}}{D_{10}} \tag{1.19}$$

$$C_c = \frac{D_{30}^2}{D_{60}D_{10}}, \tag{1.20}$$

where $C_u$ indicates the slope of the particle size distribution curve, the higher the $C_u$ value, the larger the range of particle sizes. $C_c$ indicates continuous characteristics of the slope, a well-graded soil has $1 < C_c < 3$ and $C_u > 5$.

### 1.4.2 *Method of sedimentation*

This method is based on Stokes' law which governs the velocity at which spherical particles settle in a suspension: the larger the particle, the greater is the settling velocity and vice versa.

## 1.5 Density Properties for Granular Soils

Cohesionless soils include stone, crushed stone, gravel and sand soil, which all present single grain fabric with no cohesive force and no cementation effect. Compactness is the main factor affecting the engineering properties in terms of this kind of soil. Hence, if the soil is more compact, the bearing capacity will be higher, the compressibility will be smaller, and then the stability will be much better accordingly. Since the soil particles are coarse and relative surface area is quite small, water has little influence on engineering properties in terms of this kind of soil. There is only strong bound water on the surface, and no weakly bound water.

How to express the compactness of granular soils is the key point. Besides the coefficient of uniformity $C_u$, there are usually some other ways.

### 1.5.1 *Void ratio*

Figure 1.10 shows two of the many possible ways that a system of equal-sized spheres can be packed. The dense packings represent the densest possible state for such a system. Looser systems than the simple cubic packing can be obtained by carefully constructing arches

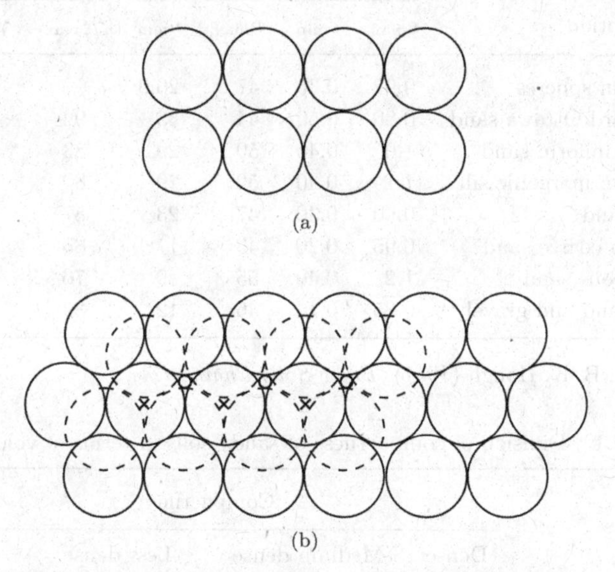


Fig. 1.10. Arrangement of uniform spheres. (a) Plan and elevation view: simple cubic packing. (b) Plan view: dense packing. Solid circles, first layer; dashed circles, second layer; ∘, location of sphere centers in third layer; face-centered cubic array; ×, location of sphere centers in third layer; close-packed hexagonal array.

*Source*: From Deresiewicz (1958).

within the packing, but the simple cubic packing is the loosest of the stable arrangements. The void ratio and porosity of these simple packings can be computed from the geometry of the packings, and the results are given in Table 1.2.

The smaller the range of particles sizes present (i.e. the more nearly uniform the soil), and the more angular the particles present, the smaller the minimum density (i.e. the greater the opportunity for building a loose arrangement of particles). The greater the range of particle sizes present, the greater the maximum density (i.e. the voids among the larger particles can be filled with smaller particles).

According to the engineering experience, Table 1.3 characterizes the density of granular soils in the basis of void ratio.

Table 1.2. Maximum and minimum densities for granular soils.

| | Void ratio | | Porosity | | Dry unit weight (pcf) | |
|---|---|---|---|---|---|---|
| Description | $e_{max}$ | $e_{min}$ | $n_{max}$ | $n_{min}$ | $\gamma_{d\,max}$ | $\gamma_{d\,min}$ |
| Uniform spheres | 0.92 | 0.35 | 47.6 | 26.0 | — | — |
| Standard Ottawa sand | 0.80 | 0.50 | 44 | 33 | 92 | 110 |
| Clean uniform sand | 1.0 | 0.40 | 50 | 29 | 83 | 118 |
| Uniform inorganic silt | 1.1 | 0.40 | 52 | 29 | 80 | 118 |
| Silty sand | 0.90 | 0.30 | 47 | 23 | 87 | 127 |
| Fine to coarse sand | 0.95 | 0.20 | 49 | 17 | 85 | 138 |
| Micaceous sand | 1.2 | 0.40 | 55 | 29 | 76 | 120 |
| Silty sand and gravel | 0.85 | 0.14 | 46 | 12 | 89 | 146 |

*Source*: B. K. Hough (1957), *Basic Soils Engineering*.

Table 1.3. Division of compactness of sandy soils in terms of void ratio.

| | Compactness | | | |
|---|---|---|---|---|
| Soil type | Dense | Medium dense | Less dense | Loose |
| Gravel, coarse or medium sand | $e < 0.6$ | $0.60 \leq e \leq 0.75$ | $0.75 < e \leq 0.85$ | $e > 0.85$ |
| Fine sand, silt | $e < 0.7$ | $0.70 \leq e \leq 0.85$ | $0.85 < e \leq 0.95$ | $e > 0.95$ |

Values of water content for natural granular soils vary from less than 0.1% for air-dry sands to more than 40% for saturated, loose sand.

### 1.5.2 *Relative density*

A useful way to characterize the density of a natural granular soil is with relative density $D_r$, defined as

$$D_r = \frac{e_{\max} - e}{e_{\max} - e_{\min}} \times 100\% = \frac{\gamma_{d\max}}{\gamma_d} \times \frac{\gamma_d - \gamma_{d\min}}{\gamma_{d\max} - \gamma_{d\min}} \times 100\%, \tag{1.21}$$

where $e_{\min}$ is the void ratio of soil in densest condition, $e_{\max}$ is the void ratio of soil in loosest condition, $e$ is the in-place void ratio, $\gamma_{d\max}$ is the dry unit weight of soil in densest condition, $\gamma_{d\min}$ is the dry unit weight of soil in loosest condition, $\gamma_d$ is the in-place dry unit weight.

According to the engineering experience, divide the compactness of sandy soils in terms of relative density $D_r$:

$$0 < D_r \leq 0.33 \text{ loose}$$

$$0.33 < D_r \leq 0.67 \text{ medium dense}$$

$$0.67 < D_r \leq 1.0 \text{ dense.}$$

A variety of tests have been proposed to measure the maximum and minimum void ratios (Kolbuszewski, 1948). The test to determine the maximum density usually involves some form of vibration. The test to determine minimum density usually involves pouring oven-dried soil into a container. Unfortunately, the details of these tests have not been entirely standardized, and values of the maximum density and minimum density for a given granular soil depend on the procedure used to determine them. By using special measures, one can obtain densities greater than the so-called maximum density. Densities considerably less than the so-called minimum density can be obtained, especially with very fine sands and silts, by slowly sedimenting the soil into water or by fluffing the soil with just a little moisture present.

Table 1.4. Standard penetration test.

| Relative density | Dense | Medium dense | Less dense | Loose |
|---|---|---|---|---|
| Penetration resistance, $N$ | $30 < N$ | $15 < N \leq 30$ | $10 < N \leq 15$ | $\leq 10$ |

### 1.5.3 *Standard penetration test*

Using void ratio $e$ and relative density $D_r$ to express the compactness of soils has many defects. Besides a lot of experimental influencing factors, it is very hard to get the intact specimens of the granular soils. Thus, many field experiments have been proposed (i.e. standard penetration test and the cone penetration test) to get the compactness of soils.

According to *Code for Design of Building Foundation* (GB 50007-2011), China, Table 1.4 presents a correlation of standard penetration resistance with relative density for sand.

The standard penetration test is a very valuable method of soil investigation. It should, however, be used only as a guide, because there are many reasons why the results are only approximate.

## 1.6 Plasticity Properties of Soils

### 1.6.1 *Plastic behavior and consistency limits*

Plasticity is a very important characteristic of fine-grained soil. In general, depending on its water content, a soil may exist in one of the liquid, plastic, semi-solid, and solid states. If the water content of a soil initially in the liquid state is gradually reduced, the state will change from liquid through plastic and semi-solid, accompanied by gradually reducing volume, until the solid state is reached. The water content at the division between the solid and the semi-solid state is the shrinkage limit $w_s$. The division between the semi-solid and plastic state is the plastic limit $w_p$. The water content indicating the division between the plastic and liquid state has been designated the liquid limit $w_L$. The liquid limit, plastic limit, and shrinkage limit are also called consistency limits (see Fig. 1.11).

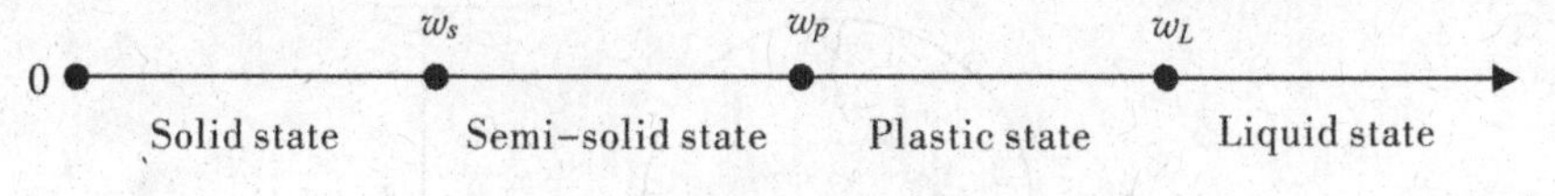


Fig. 1.11. Consistency limits.

The liquid and plastic limits of the range of water content are defined as the plasticity index $I_p$, i.e.

$$I_p = w_L - w_p. \tag{1.22}$$

For proper evaluation of a soil's plasticity properties, it has been found desirable to use both the liquid limit and plasticity index values. The natural water content ($w$) of a soil relative to the liquid and plastic limits can be represented by means of the liquidity index ($I_L$), i.e.

$$I_L = \frac{w - w_p}{w_L - w_p}. \tag{1.23}$$

A very low value for the $I_L$, or a value near zero, indicates that the water content is near the plastic limit, where experience has shown that the sensitivity will be low and the cohesive strength relatively high. As the natural water content approaches the liquid limit, the sensitivity increases.

The liquid and plastic limits are determined by means of arbitrary test procedures.

### 1.6.2 *Liquid and plastic limit tests*

#### 1.6.2.1 *Determination of the plastic limit*

There are two methods to determine the plastic limit: Rubbing method and liquid-plastic limit combined device method.

The rubbing method is relatively simple to carry out; the apparatus itself is simple, too. First, we need to make several specimens. Then, take a specimen and roll it into a ball and then we put it on a ground glass sheet to roll it evenly into a thread of soil with our palms until the diameter of the thread reaches 3 mm (Fig. 1.12). This procedure of rolling continues until the thread starts to crumble or there is no shadow mark on the glass just as the diameter of

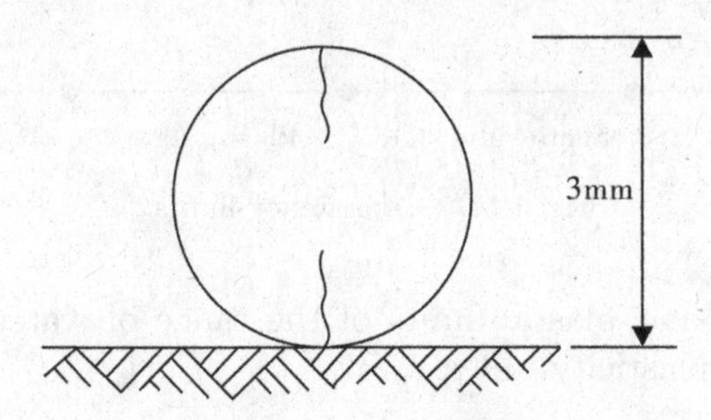


Fig. 1.12. Rubbing method.

3 mm is reached; at this point, the water content of the specimen is determined. The same process is carried out in other specimens and the average water content is stated as the plastic limit of the soil. In spite of the seeming arbitrary nature of this test procedure, an experienced technician can obtain a reasonable plastic limit result for the engineering construction.

Liquid-plastic combine tester is shown in Fig. 1.13. First, we make three kinds of specimens with different water content. Mix the three kinds of well-prepared specimens fully and put them in the soil cup. Then with the electromagnetic fall-cone test, the cone is released to penetrate the soil paste for exactly 5 s and mark down its depth, introduced in *Standard for Soil Test Method* (GB/T 50123-1999), China. The penetration procedure is repeated three times on the specimens. A log–log plot is drawn of water content/penetration depth and three points should be in a line in this plot. So the water content corresponding to a penetration of 2 mm is the value of plasticity limit.

#### 1.6.2.2 *Determination of the liquid limit*

There are two methods for the liquid limit test: liquid-plastic limit combined device method and Casagrande method, namely dish-type liquid limit device method.

Although the liquid-plastic limit combined device method is preferred, the Casagrande method of determining liquid limit is still widely used. The apparatus used is shown in Fig. 1.14 and consists basically of a metal dish which may be raised by rotating a cam and then allowed to fall through at a height of 10 mm on to a hard rubber block.

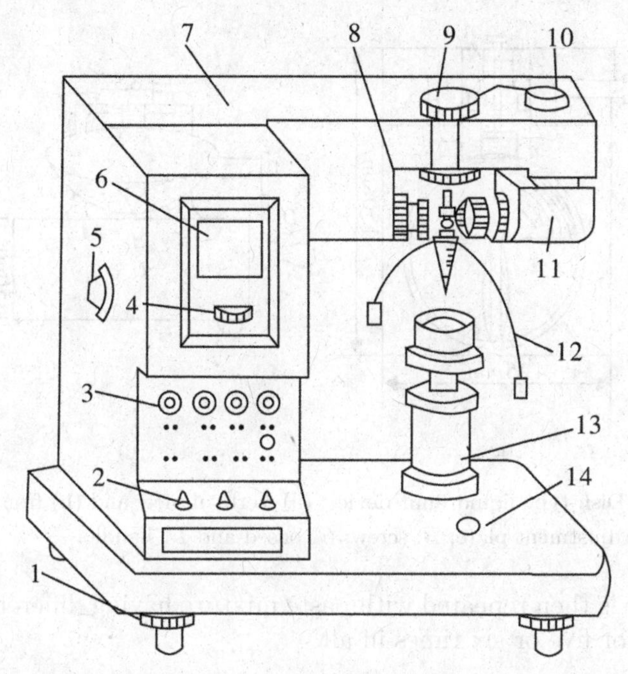


Fig. 1.13. Liquid-plastic limit combined device. (1) Horizontal adjustment screw; (2) pilot switch; (3) indicate lamp; (4) nil-line adjustment screw; (5) retroreflector adjustment screw; (6) screen; (7) chassis; (8) objective lens adjustment screw; (9) electromagnetic apparatus; (10) luminous source adjustment screw; (11) luminous source; (12) cone apparatus; (13) elevator platform, and (14) spirit bubble.

The soil is dried and mixed in the same way as in the previously described method. Some of the soil paste is placed in the dish and leveled-off parallelly with the rubber base until the depth of the specimen is 10 mm.

The standard grooving tool is then drawn through the soil paste to form the groove. By turning the handle (at 2 rvs/s), the cup is raised and dropped on to the rubber base until the lower part of the groove has closed up over a length of 13 mm. The number of blows (number of revolutions) required for this is recorded.

The dish is then refilled with the same paste mixture and the grove-closing procedure is repeated several times and an average number of blows are obtained for that mixture. After this, a small portion of the paste is taken and its water content is found. The whole

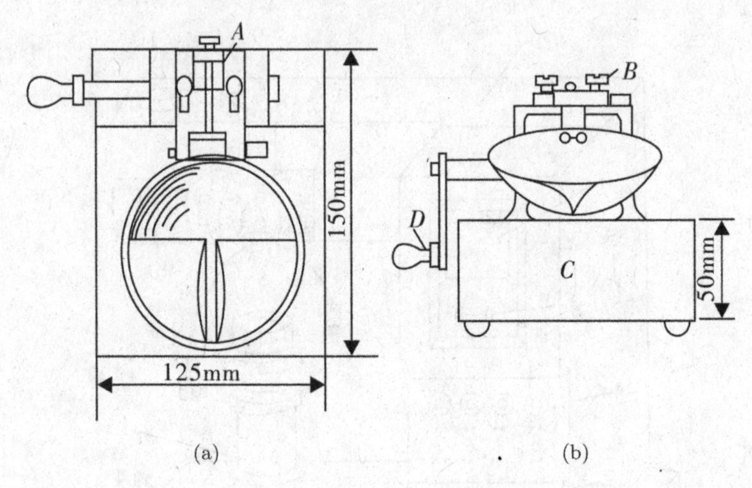


Fig. 1.14. Dish-type liquid limit device. (a) Vertical view and (b) front view.
*Notes*: $A$, Adjustment plate; $B$, screw; $C$, based and $D$, handle.

procedure is then repeated with paste mixture having different water contents for five or six times in all.

### 1.6.3 *Two important indexes and the engineering application*

#### 1.6.3.1 *Plasticity index, $I_p$*

It can be seen from Fig. 1.11 that if the value of plasticity limit $w_L$ and liquid limit $w_P$ has big difference, the range of plasticity of the soil will be wider, the weakly bound water on the surfaces of soil particles will be thicker, the specific surface area of the soil particles will be bigger and the surface adsorption capability of soil particles will be greater. Thus, plasticity index, to a large extent, indicates the amount of clay particles, colloidal particles, and clay mineral contents of the soil.

According to *Code for Design of Building Foundation* (GB 50007-2011) and *Code for Rock and Soil Classification of Railway Engineering* (TB 10077-2001), China, divide the cohesive soils on the basis of the range of plasticity index $I_p$.

$$10 < I_p \leq 17 \text{ Silty clay}$$
$$17 < I_p \text{ Clay.}$$

Table 1.5. Division of hardness of cohesive soil into five state on the basis of $I_L$.

| Degree of hardness | Solid | Stiff-plastic | Malleable | Soft-plastic | Flowing |
|---|---|---|---|---|---|
| Liquidity index | $I_L \leq 0$ | $0 < I_L \leq 0.25$ | $0.25 < I_L \leq 0.75$ | $0.75 < I_L \leq 1.0$ | $1.0 < I_L$ |

Table 1.6. Division of hardness of cohesive soil into four states on the basis of $I_L$.

| Degree of hardness | Solid | Stiff-plastic | Soft-plastic | Flowing |
|---|---|---|---|---|
| Liquidity index | $I_L \leq 0$ | $0 < I_L \leq 0.5$ | $0.5 < I_L \leq 1.0$ | $1.0 < I_L$ |

#### 1.6.3.2 *Liquidity index, $I_L$*

From Eq. (1.23) and Fig. 1.11, it can be seen that if the natural water content is less than the plastic limit ($w \leq w_p$), liquidity index will be negative ($I_L \leq 0$) and the soil will be in solid state; if the natural water content is greater than the liquid limit ($w \geq w_L$), liquidity index will be equal or greater than 1.0 ($I_L \geq 1.0$) and the soil will be in flow condition. Thus, liquidity index indicates the degree of hardness of cohesive soil.

According to *Code for Investigation of Geotechnical Engineering* (GB 50021-2001) and *Code for Design of Building Foundation* (GB 50007-2011), China, divide the cohesive soils into five states on the basis of liquidity index, $I_L$, shown in Table 1.5.

According to *Code for Rock and Soil Classification of Railway Engineering* (TB 10077-2001), China, divide the cohesive soils into four states on the basis of liquidity index, $I_L$, shown in Table 1.6.

As can be seen from Tables 1.5 and 1.6, the greater the value of liquidity index, $I_L$, the closer the liquid state of the soil mass.

## 1.7 Soil Compaction

Compaction is the process of increasing the density of a soil by packing the particles closer together with the reduction in the volume of air: there is no significant change in the volume of water of the soil. In the construction of embankments, loose soil is placed in layers ranging from 75 mm to 450 mm in thickness, each layer being compacted to a specified standard by means of rollers, vibrators, or rammers. In general, the higher the degree of compaction, the higher will the strength be and the lower the compressibility of the soil.

The degree of compaction of a soil is measured in terms of dry density, i.e. the mass of the solid only per unit volume of soil. If the bulk density of the soil is $\rho$ and the water content is $w$, then it is apparent that the dry density is given by

$$\rho_d = \frac{\rho}{1+w}. \tag{1.24}$$

The compaction characteristics of a soil can be assessed by means of the standard compaction test. The soil is compacted in a cylindrical mould using a standard compactive effort. In the test, the volume of the mould is 947 mm$^3$ and the soil (with all particles larger than 5 mm removed) is compacted by a rammer consisting of a 2.5 kg mass falling freely through 305 mm: the soil is compacted in three equal layers, each layer receiving 25 blows with the rammer. In the modified A.A.S.H.O test, the mould is the same as used in the above test, but the rammer consists of a 4.5 kg mass falling though 457 mm: the soil (with all particles larger than 40 mm removed) is compacted in five layers, each layer receiving 56 blows with the rammer.

The effectiveness of the compaction process is dependent on several factors:

(1) *The water content of soil*: The maximum dry density can't be obtained if there is just a little or a lot of water in the soil. If the water content is quite low, there is basically strong bound water in the soil and the bound water film is too thin, adding the influence of interparticle friction and attraction, which make the soil particles not quite easy to move, thus not easy to compact. If the water content is quite large, there is relatively a lot of free water in the soil, which is considered incompressible under the engineering loads. Since the free water take up a certain space, the soil with large water content is not easy to compact, either. When the water content of soil is just the optimal water content, there is some weakly bound water and no free water in the soil.

Weakly bound water film adheres to the soil particles and can move together with the soil particles, during which weakly bound water film has the lubrication effect, making soil particles easy to move, filling the voids, and becoming compact. Thus, maximum dry density can be obtained.

(2) The energy supplied by the compaction equipment (referred to as the compactive effort).
(3) The nature and type of soil (i.e. sand or clay; uniform or well graded, plastic or non-plastic).
(4) The large particles in the soil.

After compaction using one of the two standard methods, the bulk density and water content of the soil are determined and the dry density is calculated. For a given soil, the process is repeated at least five times, the water content of the sample being increased each time. Dry density is plotted against water content and a curve of the form shown in Fig. 1.15 is obtained. This curve shows that for a particular method of compaction (i.e. a particular compactive effort), there is a particular value of water content, known as the optimum water content ($w_{op}$), at which a maximum value of dry density is obtained. At low values of water content, most soils tend to be stiff and are difficult to compact. As the water content

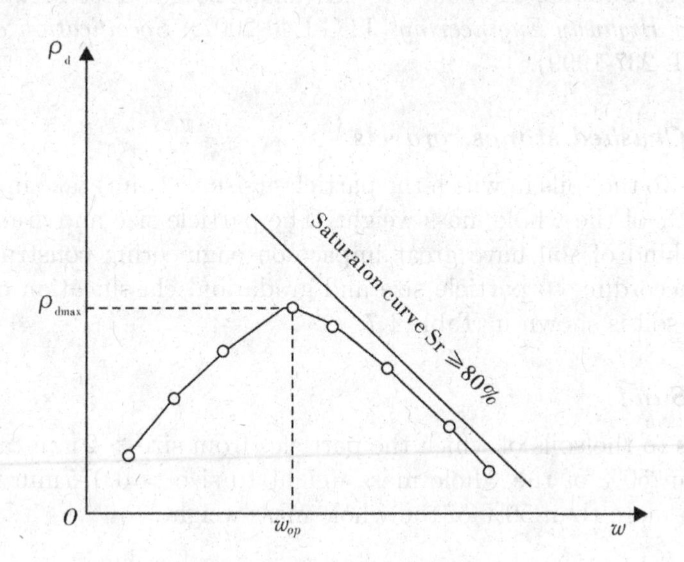


Fig. 1.15. Soil compaction curve.

increases, the soil becomes more workable, facilitating compaction and resulting in higher dry densities, which is very beneficial for engineering construction. At high water contents, however, the dry density decreases with increasing water content, and an increasing proportion of the soil volume is occupied by water. So, dry density is quite an important quality inspection index for compaction.

## 1.8 Engineering Classification of Soils

The direct approach to the solution of a soil engineering problem consists of first measuring the soil property needed and then employing this measured value in some rational expression to determine the answer to the problem.

Measuring fundamental soil properties like permeability, compressibility, and strength can be difficult, time consuming, and expensive. In many soil engineering problems, such as pavement design, there are no rational expressions available for the analysis for the solution. For these reasons, sorting soils into groups showing similar behavior may be very helpful. Such sorting is soil classification.

In China, several kinds of classification standards apply: *Standard for Engineering Classification of Soil* (GB/T 50145-2007); *Code for Design of Building Foundation* (GB 50007-2011); *Test Methods of Soils for Highway Engineering* (JTG E40-2007); *Specification of Soil Test* (SL 237-1999).

### 1.8.1 *Crushed stones, gravels*

It refers to the soils in which the particles (size > 2 mm) take up more than 50% of the whole mass weight. The particle size and gradation of this kind of soil have great impact on engineering construction. Thus, according to particle size and gradation, classification of this kind of soil is shown in Table 1.7.

### 1.8.2 *Sand*

It refers to the soils of which the particles from size > 2 mm take up less than 50% of the whole mass weight to size > 0.075 mm which take up more than 50% of the whole mass weight.

Table 1.7. Classification of crushed stones and gravels.

| Soil type | Soil particle shape | Gradation |
|---|---|---|
| Boulder | Round forms dominant | Particles (size > 200 mm) take up more than 50% of the whole mass weight. |
| Block stone | Angular forms dominant | |
| Cobble | Round forms dominant | Particles (size > 20 mm) take up more than 50% of the whole mass weight. |
| Crushed stone | Angular forms dominant | |
| Round gravel | Round forms dominant | Particles (size > 2 mm) take up more than 50% of the whole mass weight. |
| Angular gravel | Angular forms dominant | |

Table 1.8. Classification of sand in the basis of gradation.

| Soil type | Gradation |
|---|---|
| Gravelly sand | Particles (size > 2 mm) take up 25–50% of the whole mass weight. |
| Coarse sand | Particles (size > 0.5 mm) take up 50% of the whole mass weight. |
| Medium sand | Particles (size > 0.25 mm) take up 50% of the whole mass weight. |
| Fine sand | Particles (size > 0.075 mm) take up 85% of the whole mass weight. |
| Silty sand | Particles (size > 0.075 mm) take up 50% of the whole mass weight. |

For this kind of soil, the classification code is similar to the crushed stones and gravels, as shown in Table 1.8.

According to the void ratio, relative density, and standard penetration test, this kind of soil can also be classified, as shown in Section 1.5.

### 1.8.3 *Silty soil*

It refers to the soils in which particles (size $> 0.075\,\mathrm{mm}$) take up less than 50% of the whole mass weight and the plasticity index $I_p \leq 10$.

### 1.8.4 *Clay*

In the basis of plasticity index $I_p$, clay can be divided as shown in Section 1.6.3.

According to the degree of hardness, clay can be divided as shown in Tables 1.5 and 1.6.

## Exercises

1.1 Explain the concept of plasticity index and liquidity index and its application in engineering.

1.2 What is particle size distribution curve and how to use it in engineering?

1.3 How to appraise the compactness of granular soils?

1.4 What are maximum dry density and optimum water content of soil?

1.5 A saturated clay specimen has the water content of 36.0%, and particle specific gravity of 2.70. Determine the void ratio and dry density.

1.6 The volume of a soil specimen is $60\,\mathrm{cm}^3$, and its mass is 108 g. After being dried, the mass of the sample is 96.43 g. The value of $d_s$ is 2.7. Calculate wet density, dry density, water content, porosity, and the degree of saturation.

1.7 A soil specimen has a water content $w_1$ of 12% and unit weight $\gamma_1$ of $19.0\,\mathrm{kN/m}^3$. Keep void ratio constant and increase water content to $w_2 = 22\%$. How much water will be infused into $1\,\mathrm{m}^3$ soil?

1.8 There is a sample which is got from a certain natural sand column. The water content measured by experiment is 11%, bulk density $\rho = 1.70\,\mathrm{g/cm}^3$, the minimum dry density is $1.41\,\mathrm{g/cm}^3$, the maximum dry density is $1.75\,\mathrm{g/cm}^3$. Determine the sandy soil's dense degree.

1.9 Calculate the dry unit weight, the saturated unit weight, and the buoyant unit weight of a soil having a void ratio of 0.70 and a value of $d_s$ of 2.72. Calculate also the unit weight and water content at a degree of saturation of 75%.

## Bibliography

J. Atkinson (1993). *An Introduction to the Mechanics of Soils and Foundations.* McGraw-Hill, England.

R. F. Craig (1998). *Soil Mechanics.* E and FN Spon, London and New York.

H. Deresciewice (1958). *Mechanics of Granular Matter. Advances in Applied Mechanics*, Vol. 5. Academic Press, New York.

B. K. Hough (1957). *Basic Soils Engineering.* The Ronald Press Company, New York.

J. J. Kolbuszewski (1948). An Experimental Study of the Maximum and Minimum Porosities of Sands, *Proc. 2nd Inter. Conf. Soil Mech. Found. Eng.* (Rotterdam), Vol. I, p. 158.

H. Liao (2018). *Soil Mechanics* (Third Edition). Higher Education Press, Beijing.

Ministry of Housing and Urban-Rural Construction of the People's Republic of China, GB 50007-2011 (2012). *Code for Design of Building Foundation.* China Building Industry Press, Beijing.

Ministry of Water Resources of the People's Republic of China, GB/T 50123-1999 (2000). *Standard for Soil Test Method.* China Planning Press, Beijing.

Ministry of Construction of the People's Republic of China, GB 50021-2001 (2009). *Code for Investigation of Geotechnical Engineering* (The 2009 Revised Edition). China Building Industry Press, Beijing.

Ministry of Railways of the People's Republic of China, TB 10077-2001 (2001). *Code for Rock and Soil Classification of Railway Engineering.* China Railway Press, Beijing.

Ministry of Water Resources of the People's Republic of China, GB/T 50145-2007 (2008). *Standard for Engineering Classification of Soil.* China Planning Press, Beijing.

Ministry of Communications of the People's Republic of China, JTG E40-2007 (2007). *Test Methods of Soils for Highway Engineering.* China Communications Press, Beijing.

Ministry of Water Resources of the People's Republic of China, SL 237-1999 (2003). *Specification of Soil Test.* China Standards Press, Beijing.

Soil Mechanics Work Team at Hohai University (2004). *Soil Mechanics.* China Communication Press, Beijing.

K. von Terzaghi (1943). *Theoretical Soil Mechanics.* John Wiley and Sons, New York.

T. William Lambe and Robert V. Whitman (1969). *Soil Mechanics.* John Wiley and Sons, New York.

S. Zhao and H. Liao (2009). *Civil Engineering Geology.* Science Press, Beijing.

## Chapter 2

# Permeability of Soil and Seepage Force

### Guideline

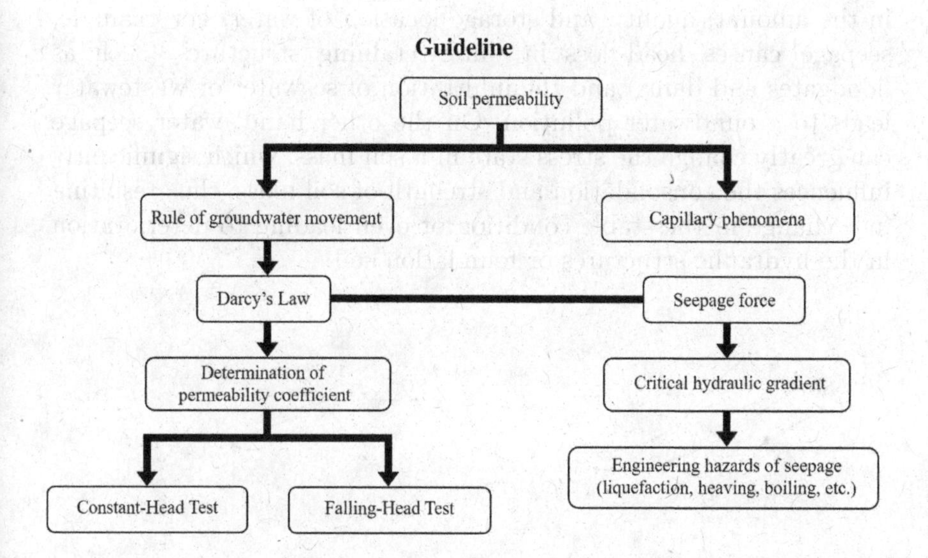

## 2.1 Introduction

Any given mass of soil consists of solid particles of various sizes together with interconnected void spaces within. Soil is directly used as a material in hydraulic structures, where water can flow under the effect of water-level difference from a side of high energy to a side of low energy through the continuous voids that are present in the soil, which is quite common in hill slopes, dams, ground, and foundation pits, etc. (see Fig. 2.1).

From the above-mentioned effect of water-level difference, permeability can be defined as the property of a soil that allows the seepage of fluids through its interconnected void spaces.

Water seepage in soil can cause complex interactions between water and soil, which lead to various problems in construction engineering. On the one hand, a lot of problems arise due to a change in the amount, quality, and storage location of water. For example, seepage causes head loss in water-retaining structures (such as floodgates and dams) and the infiltration of seawater or wastewater leads to groundwater pollution. On the other hand, water seepage can greatly change the stress state in a soil mass, which significantly influences the consolidation and strength of soil mass, thus resulting in a change in the stable condition or even leading to deterioration in the hydraulic structures or foundation soil.

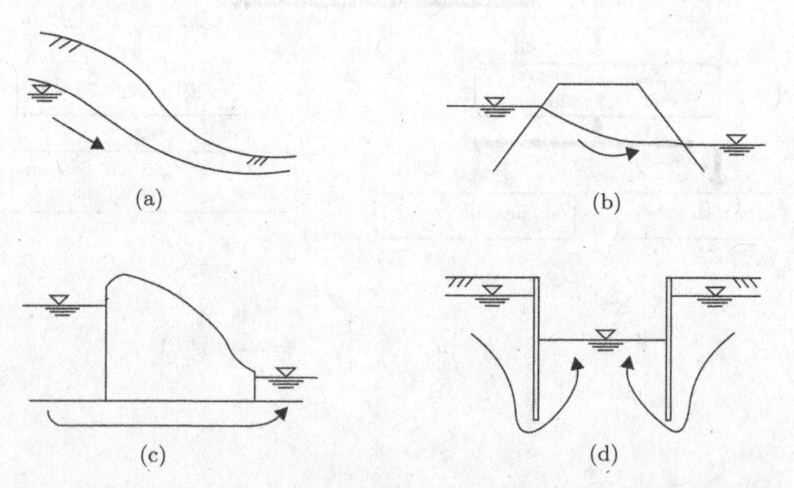


Fig. 2.1. Seepage phenomena. (a) Seepage in the slope, (b) seepage in the earth dam, (c) seepage in the ground, and (d) seepage in the foundation pit.

It can be seen from the above examples that understanding the influence of the fluid on the soil through which it flows is significant. This chapter is devoted to the study of the basic parameters and the relevant engineering problems involved in the flow of water through soils.

## 2.2 Capillary Phenomena

There are numerous evidences that a liquid surface resists tensile forces because of the attraction between adjacent molecules in the surface. This attraction is measured by surface tension, a constant property of any pure liquid that is in contact with another liquid or with a gas at a given temperature. An example of these evidences is the fact that in fine-grained soils (the average diameter of voids is 0.002–0.5 mm) water is capable of rising to a considerable height above the water table and remaining there permanently. This kind of water is called capillary water. This phenomenon is commonly referred to as capillarity.

The soil layer wetted by capillarity is called capillary water zone. Based on the formation and distribution conditions, capillary water zone can be divided into three parts — normal capillary water zone, capillary network zone, and capillary suspension zone.

(1) **Normal capillary zone (zone of capillary saturation):** This part lies at the bottom of the capillary water zone. The capillary water is mainly formed due to a direct increase in the water table. All the voids are almost filled with capillary water. The normal capillary zone makes changes accordingly depending on the rise and fall of underground water level.
(2) **Capillary network zone:** This zone lies in the middle of the capillary water zone. With a quick fall in groundwater level, the level of this zone also quickly dips, but some capillary water in the thinner capillary voids can't make quick movement and still stays in the voids. As a result of the drop in the level of the capillary water in some larger voids, air bubbles are created in the voids, causing a netted distribution in the capillary water level. The water particles in this zone can move due to the effect of surface tension and gravity.

(3) **Capillary suspension zone:** This zone lies on the top of the entire capillary water zone. The capillary water in this part is formed by the infiltration of surface water. The water particles are suspended in the soil particles and have no connection with the capillary water in the middle and bottom regions. This part of suspended capillary water can make downward movement due to the effect of gravity by atmospheric precipitation.

The above-mentioned three kinds of capillary water zones don't necessarily exist at the same time, which depends on the local hydrogeological conditions. When the underground water level is high, maybe only normal capillary zone exists; on the contrary, when the underground water level is low, the three kinds of capillary water can exist at the same time.

The capillary water in soil has a significant impact on construction engineering, some of which include the following:

(1) Since capillary water is above the surface of the free water, it leads to the change of dry and wet states in roadbed. Hence, the rise of capillary water is one of the main factors that results in the damage of the roadbed.
(2) For buildings, the rise of capillary water can lead to a high level of moisture in the basement, so the damp course becomes an absolute necessity at the top of the basement.
(3) The rise of capillary water can lead to soil paludification and salinization, affecting construction engineering and agricultural economic development.

However, due to the temporary cohesion created as a result of the capillary phenomena in the wet sand, it is possible to obtain an upright slope of 2–3 m height thus making the creation of sand sculptures feasible in the coastal region.

## 2.3 Groundwater Movement and Darcy's Law

### 2.3.1 *Laminar flow and turbulent flow*

#### 2.3.1.1 *Laminar flow*

Due to the groundwater seepage, the flow lines formed by the water particles are parallel in all places. The soil presents at the bottom cannot float upward; leaves or other lightweight objects that float on the water surface are retained on the surface and cannot be pulled downward. Through a certain space, water flows smoothly with uniform velocity and the flow velocity in the midst of water cross-section is higher, but is smaller on both sides. Based on the flow characteristics stated above, the flow is commonly referred to as laminar flow. The flow velocity, flow direction, water level, hydraulic pressure, and a few other motion indexes at any point in this seepage field don't change over time, which is called steady flow movement.

#### 2.3.1.2 *Turbulent flow*

Compared with laminar flow, in this kind of groundwater seepage, the flow lines cross each other. The flow presents twisted, mixed, and irregular movements and exists as hydraulic drops and whirlpools. Based on the flow characteristics stated above, this flow is commonly referred to as turbulent flow. All the motion indexes at any point in this seepage field change over time, which is called unsteady flow movement.

### **2.3.2 *Darcy's law***

The voids in a soil (sand, clay) are generally quite small. Although the real seepage of water through the small void spaces in a soil is irregular, the flow can be regarded as laminar flow, because the movement of water through the void spaces is very slow. Under the condition of laminar flow, the discharge velocity and energy loss obey the linear seepage relationship, which was obtained by Darcy in 1856.

In order to obtain a fundamental relation for the quantity of seepage through a soil mass under a given condition, the case shown in Fig. 2.2 is considered. The cross-sectional area of the soil is equal to $A$ and the rate of seepage is $Q$. Darcy experimentally found that $Q$ was proportional to $i$, that is

$$Q = k \cdot A \cdot i. \tag{2.1}$$

Note that $A$ is the cross-sectional area of the soil perpendicular to the direction of flow.

The hydraulic gradient $i$ can be given by

$$i = \frac{H_1 - H_2}{L} = \frac{\Delta H}{L},$$

where $L$ is the distance between the two piezometric tubes.

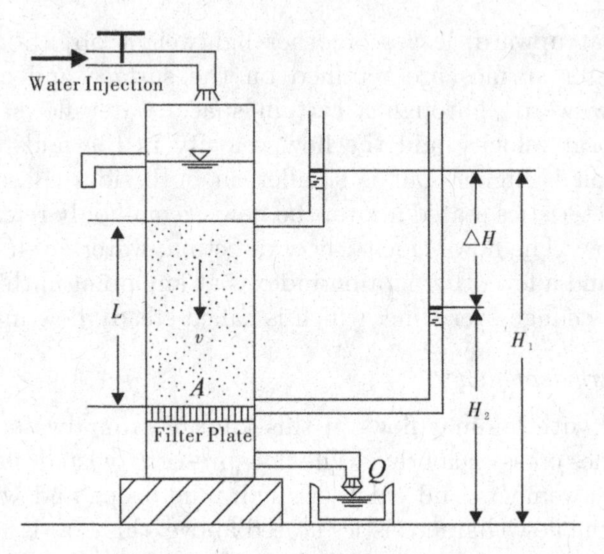


Fig. 2.2. Permeability test device.

A further consideration of the velocity at which a drop of water moves as it flows through soil helps to understand fluid flow. Equation (2.1) can be rewritten as

$$\frac{Q}{A} = k \cdot i = v.$$

Therefore, for sandy soils, Darcy published a linear relation between the discharge velocity and the hydraulic gradient

$$v = ki, \tag{2.2}$$

where $v$ is the discharge velocity (cm/s), which is not the actual velocity of seepage through soil, and the velocity $v$ given by Eq. (2.2) is the discharge velocity calculated on the basis of the gross cross-sectional area; $i$ is the hydraulic gradient, the water-level difference per unit length along the flow direction; $k$ is the coefficient of permeability, cm/s, a measure of the resistance of the soil to flow of water.

Darcy's law given by Eq. (2.2), $v = ki$, is valid for laminar flow through the void spaces. Several studies have been conducted to investigate the range over which Darcy's law is valid, and Reynolds number was obtained as a result of an excellent summary of these works. For flow through soils, Reynolds number $R_n$ can be given by the relation

$$R_n = \frac{vD\rho}{\mu}, \tag{2.3}$$

where $v$ is the discharge velocity (cm/s), $D$ is the average diameter of the soil particle (cm), $\rho$ is the density of the fluid (g/cm$^3$), $\mu$ is the coefficient of viscosity (g/(cm s)).

For laminar flow conditions in soils, experimental results showed that

$$R_n = \frac{vD\rho}{\mu} \leq 1.$$

With coarse sand, assume $D = 0.45\,\text{m}$ and $k \approx 100D^2 = 100(0.045)^2 = 0.203\,\text{cm/s}$.

Assume $i = 1$, then we get $v = ki = 0.203\,\text{cm/s}$. Also, let $\rho_{\text{water}} \approx 1\,\text{g/cm}^3$ and $\mu_{20^\circ\text{C}} = (10^{-5})(981)\,\text{g/(cm s)}$. Hence, we obtain

$$R_n = \frac{(0.203)(0.045)(1)}{(10^{-5})(981)} = 0.931 < 1.$$

From the above calculations, we can conclude that, flow of water through all types of soil (sand, silt, and clay) is laminar and that Darcy's law is valid. With coarse sand, gravels, and boulders, turbulent flow of water can be expected.

Darcy's law as defined by Eq. (2.2) implies that the discharge velocity bears a linear relation with the hydraulic gradient for sand (Fig. 2.3(a)).

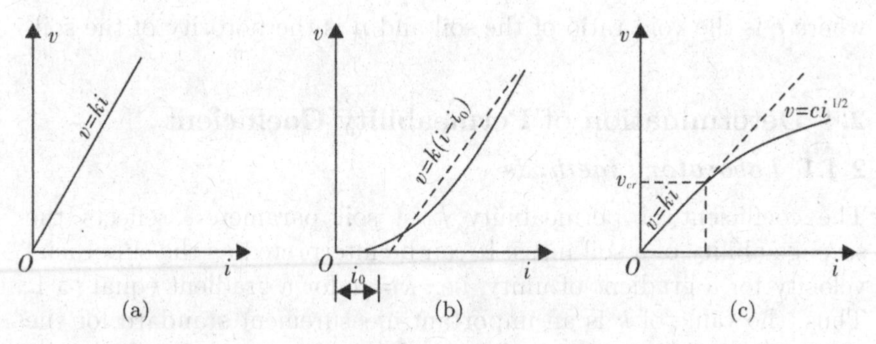


Fig. 2.3. Discharge velocity–hydraulic gradient relationship of soil. (a) Sand, (b) clay, and (c) gravel.

Hansbo (1960) reported the results of four undisturbed natural clays (Fig. 2.3(b)). On the basis of his results, it is seen that

$$v = k(i - i_0), \quad i \geq i_0,$$

and

$$v = ki^n, \quad i < i_0.$$

The value of $n$ for the four Swedish clays was about 1.6. There are several studies, however, that refute the preceding conclusion.

For gravel or other coarse-grained soils, the discharge velocity bears a linear relation with the hydraulic gradient just when the hydraulic gradient is small; when the hydraulic gradient is quite large, turbulent flow of water can be expected. Discharge velocity and hydraulic gradient will not present a linear relationship (Fig. 2.3(c)).

It must be pointed out that the velocity $v$ given by Eq. (2.3) is the discharge velocity calculated on the basis of the gross cross-sectional area. Since water can flow only through the interconnected pore spaces, the actual velocity of seepage through soil is $v'$. Assuming the rate of seepage is $Q$, the cross-sectional area of the soil is equal to $A$. Hence, the actual cross-water area

$$A' = nA.$$

According to the continuity of water, $Q = vA = v'A'$, then we have

$$v = v' \times \frac{A'}{A} = v'n = v'\frac{e}{1+e}, \tag{2.4}$$

where $e$ is the void ratio of the soil and $n$ is the porosity of the soil.

## 2.4 Determination of Permeability Coefficient

### 2.4.1 *Laboratory methods*

The coefficient of permeability $k$, a soil parameter, reflects the seepage ability of a soil mass. It can be interpreted as the superficial velocity for a gradient of unity, i.e., $k = v$ for a gradient equal to 1. Thus, the value of $k$ is an important measurement standard for the soil seepage ability. It can't be calculated directly and needs to be measured in laboratory tests.

The two most common laboratory methods for determining the coefficient of permeability of soils are constant-head test and falling-head test. In both cases, water flows through a soil sample and the rates of flow and the hydraulic gradients are measured.

It needs to be noted that the values of the coefficient of permeability measured in laboratory permeameter tests are often highly inaccurate, for a variety of reasons such as anisotropy (i.e., values of $k$ different for horizontal and vertical flow) and small samples being unrepresentative of large volumes of soil in the ground, but in practice, the values of $k$ measured from *in situ* tests are much better.

#### 2.4.1.1 *Constant-head test*

The constant-head test is suitable for more permeable granular materials. The basic laboratory test arrangement is shown in Fig. 2.4. The soil specimen is placed inside a cylindrical mold, and the constant-head loss $h$ of water flowing through the soil is maintained by adjusting the supply. The outflow water is collected in a measuring cylinder, and the duration of the collection period is recorded. From Darcy's law, the total quantity of flow $Q$ in time $t$ can be given by

$$Q = vAt = kiAt,$$

where $A$ is the area of cross-section of the specimen. However, $i = h/L$, where $L$ is the length of the specimen, so $Q = k(h/L)At$.

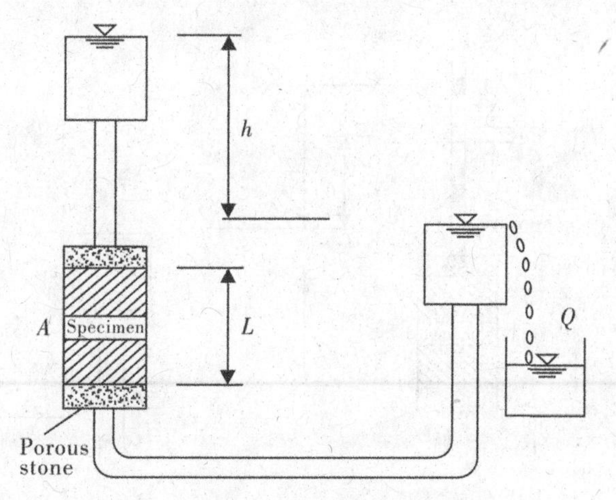


Fig. 2.4. Constant-head laboratory permeability test.

Rearranging gives

$$k = \frac{QL}{Aht}. \tag{2.5}$$

Once all the variables on the right-hand side of Eq. (2.5) have been determined from the test, the coefficient of permeability of the soil can be calculated.

#### 2.4.1.2 *Falling-head test*

The falling-head permeability test is more suitable for fine-grained soils. Figure 2.5 shows the general laboratory arrangement for the test. The soil specimen is placed inside a tube, and a standpipe is attached to the top of the specimen. Water that comes from the standpipe flows through the specimen. The initial head difference $h_1$ at time $t = t_1$ is recorded, and then water is allowed to flow through the soil such that the final head difference at time $t = t_2$ is $h_2$.

The rate of flow through the soil is

$$Q = kiA = k\frac{h}{L}A = -a\frac{dh}{dt}, \tag{2.6}$$

where $h$ is the head difference at any time $t$, $A$ is the area of the specimen, $a$ is the area of the standpipe, $L$ is the length of the specimen.

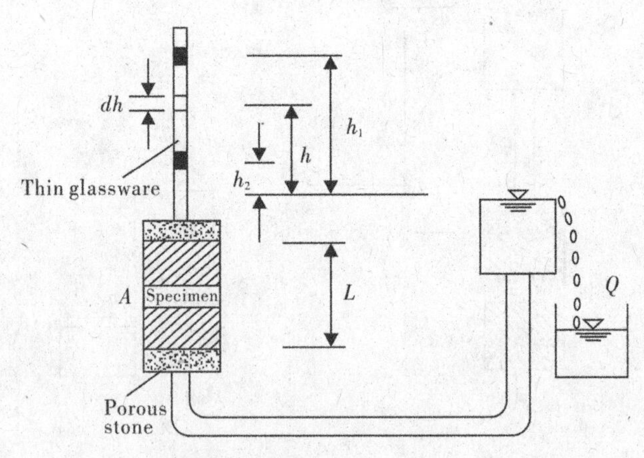


Fig. 2.5. Falling-head laboratory permeability test.

From Eq. (2.6),

$$k = \frac{aL}{A(t_2 - t_1)} \ln \frac{h_1}{h_2}, \tag{2.7}$$

or

$$k = 2.3\frac{aL}{A(t_2 - t_1)} \lg \frac{h_1}{h_2}. \tag{2.8}$$

The values of $a$, $L$, $A$, $t_1$, $t_2$, $h_1$, and $h_2$ can be determined from the test, and the coefficient of the permeability $k$ for a soil can then be calculated from Eq. (2.7) or Eq. (2.8).

There are several factors affecting the coefficient of permeability, such as soil types, gradation, void ratio, and water temperature. Hence, in order to make a precise calculation of the coefficient of permeability, we must try to keep the original state of a soil and eliminate any possible influences of artificial interference. The reference values of the coefficient of permeability of some soil types are listed in Table 2.1.

### 2.4.2 *Effective coefficient of permeability for stratified soils*

In general, natural soil deposits are stratified. If the stratification is continuous, the effective coefficients of permeability for flow in the horizontal and vertical directions can be readily calculated.

Table 2.1. Reference values of the coefficient of permeability.

| Basic soil type | Coefficient of permeability k/(cm s$^{-1}$) | Degree of permeability |
|---|---|---|
| Pure gravels | $> 10^{-1}$ | High |
| Mix of pure gravels and other types of gravels | $10^{-3} - 10^{-1}$ | Middle |
| Finest sand | $10^{-5} - 10^{-3}$ | Low |
| Mix of silt, sand, and clay | $10^{-7} - 10^{-5}$ | Very low |
| Clay | $< 10^{-7}$ | Almost impermeable |

#### 2.4.2.1 *Flow in the horizontal direction*

Figure 2.6 shows several layers of soil with horizontal stratification. Owing to fabric anisotropy, the coefficient of permeability of each soil layer may vary depending on the direction of flow. Therefore, let us assume that $k_1, k_2, \ldots, k_n$ are the coefficients of permeability for layers $1, 2, \ldots, n$, respectively, for flow in the horizontal direction.

Considering unit width of the soil layers as shown in Fig. 2.6, the rate of seepage in the horizontal direction can be given by

$$Q_x = Q_{1x} + Q_{2x} + \cdots + Q_{nx} = \sum_{i=1}^{n} Q_{ix}, \tag{2.9}$$

where $Q$ is the flow rate through the combined stratified soil layers and $Q_{1x}, Q_{2x}, \ldots, Q_{nx}$ are the rates of flow through soil layers $1, 2, \ldots, n$, respectively. Note that for flow in the horizontal direction (which is the direction of stratification of the soil layers), the

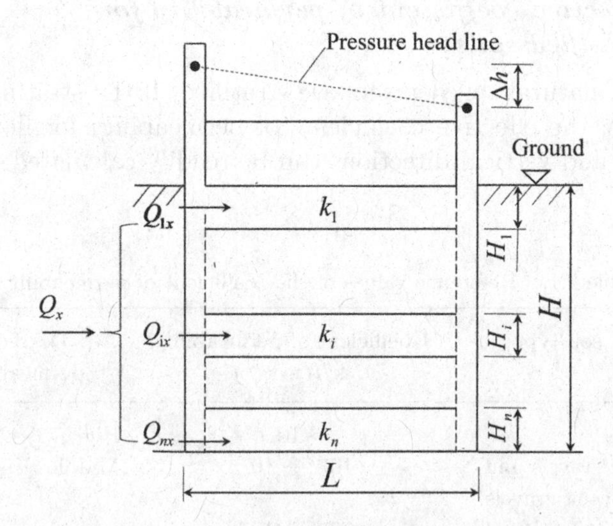


Fig. 2.6. Flow in horizontal direction in stratified soil.

hydraulic gradient is the same for all layers ($i_1 = i_2 = \cdots i_n = i = \frac{\Delta h}{L}$). So,

$$\begin{aligned} Q_{1x} &= k_1 i H_1 \\ Q_{2x} &= k_2 i H_2 \\ &\cdots \\ Q_{nx} &= k_n i H_n, \end{aligned} \tag{2.10}$$

where $i$ is the hydraulic gradient, $H_1, H_2, \ldots, H_n$ are the thicknesses of layers 1, 2, ..., $n$, respectively, and

$$Q_x = k_x i H, \tag{2.11}$$

where $k_x$ is the effective coefficient of permeability for flow in the horizontal direction

$$H = H_1 + H_2 + \cdots + H_n.$$

Substituting Eqs. (2.10) and (2.11) into Eq. (2.9) yields

$$k_x H = k_1 H_1 + k_2 H_2 + \cdots + k_n H_n.$$

Hence,

$$k_x = \frac{1}{H}(k_1 H_1 + k_2 H_2 + \cdots k_n H_n) = \frac{1}{H}\sum_{i=1}^{n} k_i H_i. \tag{2.12}$$

From Eq. (2.12), for flow in the horizontal direction, it can be seen that if the thickness of each soil layer is close and the coefficient of permeability is quite different, then the value of $k_x$ depends on the most permeable soil layer ($k'$ and $H'$ are the permeability coefficient and thickness of the most permeable soil layer, respectively). Thus, the value of $k_x$ is approximately equal to $k'H'/H$.

### 2.4.2.2 *Flow in the vertical direction*

For flow in the vertical direction for the soil layers shown in Fig. 2.7,

$$Q_y = Q_{1y} = Q_{2y} = \cdots = Q_{ny} \tag{2.13}$$

where $Q_{1y}, Q_{2y}, \ldots, Q_{ny}$ are the discharge velocities in layers $1, 2, \ldots, n$, respectively.

Note that for flow in the vertical direction, the head loss of each soil layer is $\Delta h_i$ and the hydraulic gradient $i_i$ is $\Delta h_i/H_i$, so

$$Q_{iy} = k_1 i_1 = k_2 i_2 = \cdots = k_n i_n = k_i \frac{\Delta h_i}{H_i} A, \tag{2.14}$$

where $k_1, k_2, \ldots, k_n$ are the coefficients of permeability for layers $1, 2, \ldots, n$, respectively, for flow in the vertical direction, $i_1, i_2, \ldots, i_n$ are the hydraulic gradients in soil layers $1, 2, \ldots, n$, respectively, $A$ is the area of cross-section of the soil perpendicular to the direction of flow.

The head loss of the entire soil mass $h$ is $\sum \Delta h_i$, the overall average hydraulic gradient $i$ is $h/H$, so

$$Q_y = k_y \frac{h}{H} A, \tag{2.15}$$

where $k_y$ is the effective coefficient of permeability for flow in the vertical direction.

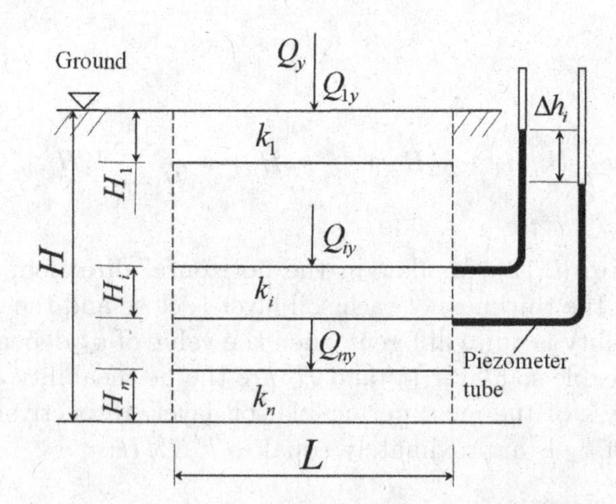


Fig. 2.7. Flow in vertical direction in stratified soil.

Substitution of Eqs. (2.14) and (2.15) into Eq. (2.13) yields

$$k_y = \frac{H}{\sum_{i=1}^{n} \left(\frac{H_i}{k_i}\right)}. \tag{2.16}$$

From Eq. (2.16), for flow in the vertical direction, it can be seen that if the thickness of each soil layer is similar and the coefficient of permeability is quite different, then the value of $k_y$ depends on the most impermeable soil layer ($k''$ and $H''$ are the permeability coefficient and thickness of the most impermeable soil layer, respectively). Thus, the value of $k_y$ is approximately equal to $k''H/H''$.

### 2.4.3 *Factors affecting the coefficient of permeability*

The coefficient of permeability depends on several factors, most of which are listed as follows:

(1) **Shape and size of the soil particles:** Permeability increases when the soil particles are much coarser and rounded.
(2) **Void ratio:** Permeability increases with an increase in void ratio. In soil with good gradation, fine soil particles fill in coarse soil particles, causing the void ratio to decrease and eventually leading to a decrease in permeability.
(3) **Degree of saturation:** Permeability increases with an increase in the degree of saturation.
(4) **Composition of soil particles:** It is not an important factor for sands and silts. However, for soils with clay minerals, this is one of the most important factors. Permeability depends on the thickness of water surrounding the soil particles, which is a function of the cation exchange capacity, valency of the cations, and so forth. Other factors remaining the same, the coefficient of permeability decreases with increasing thickness of the diffuse double layer.
(5) **Soil structure:** Fine-grained soils with a flocculated structure have a higher coefficient of permeability than those with a dispersed structure.
(6) **Viscosity of the fluid.**
(7) **Density and concentration of the fluid.**

## 2.5 Flow Nets

### 2.5.1 *Definition*

A set of flow lines and equipotential lines is called a flow net. A flow line is the route through which a water particle travels. An equipotential line is a line joining the points that show the same piezometric elevation. Figure 2.8 shows an example of a flow net for a dam. The permeable layer is isotropic with respect to the coefficient of permeability, i.e., $k_x = k_z = k$. Note that the solid lines in Fig. 2.8 are the flow lines and the broken lines are the equipotential lines. In drawing a flow net, the boundary conditions must be kept in mind.

It must be remembered that the flow lines intersect with the equipotential lines at right angles. The flow and equipotential lines are usually drawn in such a way that the flow elements are approximately squares.

### 2.5.2 *Calculation of seepage from a flow net under a hydraulic structure*

Soil seepage can be qualitatively determined by the flow net. As we can see from Figure 2.8 that near the draining prism there are the densest flow lines, showing that the values of hydraulic gradient and seepage velocity are also the largest in this region. Accordingly, the flow lines far away from the seepage prism are quite loose with corresponding smaller values of hydraulic gradient and seepage velocity.

If the quantity of flow between the two consecutive flow lines is the same and the potential energy difference between the two

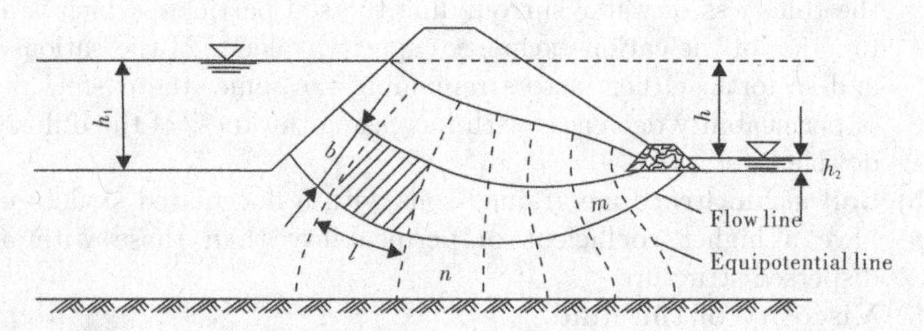


Fig. 2.8. A flow net around a dam.

consecutive equipotential lines is also the same, we can obtain the following results:

(1) The flow line density indicates the different intensities of groundwater flow; the denser the flow lines are, the greater the flow intensity is.
(2) The equipotential line density indicates the rate of hydraulic gradient change; the denser the equipotential lines are, the greater the hydraulic gradient is.

Suppose that flow elements are squares ($b = l$) and assume $dq$ indicates the flow through the flow channel per unit length of the hydraulic structure (i.e., perpendicular to the section shown). Then, according to Darcy's law, we have

$$dq = -b \cdot v = -b \cdot ki = -b \cdot k\frac{dh}{l} \approx -kdh. \tag{2.17}$$

Suppose that there are $n$ potential drops and $m$ flow channels in a flow net, then the rate of seepage per unit length of the hydraulic structure is

$$q = n \cdot dq \doteq n \cdot kdh = n \cdot k\frac{h}{m} = \frac{n}{m} \cdot kh. \tag{2.18}$$

Besides, if the values of flow net are given, the hydraulic head, hydraulic gradient, seepage quantity, pore water pressure, and seepage force at every point in the seepage field can be calculated.

## 2.6 Seepage Force

### 2.6.1 *Conception of seepage force*

Flow of water through a soil mass results in a certain force being exerted on the soil itself, which is defined as the seepage force, it is a body force and is indicated by $G_d$ (kN/m$^3$). The resistance to the seepage water for soil particles $T$ is also known as body force and $T = -G_d$.

To evaluate the seepage force per unit volume of a soil, a water column $BA$ is considered where the length and section area are represented as $L$ and $A$, respectively, as shown in Fig. 2.9. The weight of the water column is $\gamma_w LA$, the hydrostatic force on the

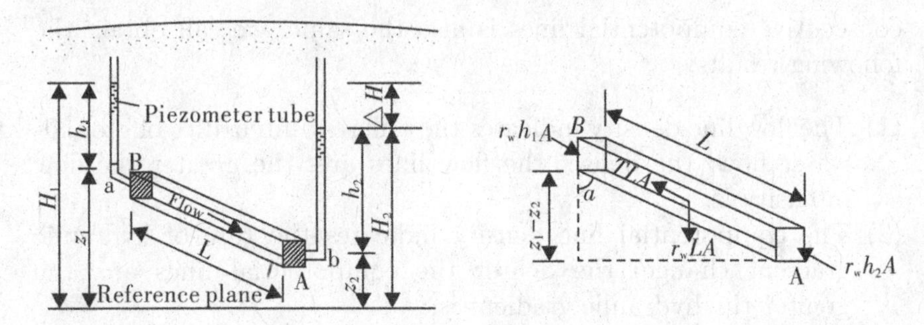


Fig. 2.9. Force analysis on the water column.

side $B$ is $\gamma_w h_1 A$, the hydrostatic force on the side $A$ is $\gamma_w h_2 A$, and the resistance of the seepage water from soil particles is $TAL$. For equilibrium,

$$\gamma_w h_1 A + \gamma_w LA \cos\alpha - \gamma_w h_2 A - TAL = 0, \tag{2.19}$$

$$\cos\alpha = \frac{z_1 - z_2}{L}. \tag{2.20}$$

However, $z_1 + h_1 = H_1$, $z_2 + h_2 = H_2$, $i = \frac{H_1 - H_2}{L}$, so from Eqs. (2.19) and (2.20), we get

$$T = \gamma_w i. \tag{2.21}$$

Hence, the seepage force $G_d$ is given as

$$G_d = \gamma_w i. \tag{2.22}$$

The direction of seepage force is the same as the flow direction. From Eq. (2.22), $G_d$ is the body force.

Sometimes $G_d$ is expressed by surface force or stress as follows:

$$u = G_d L = \gamma_w i L, \tag{2.23}$$

where $L$ is the length of flow (m) and $u$ is the seepage force or stress (Pa), at the exit of seepage, $u = 0$.

With the change in flow direction, the seepage force will have different effects on the soil, as shown in Fig. 2.10. When upward seepage occurs, the direction of seepage force and gravity is just opposite, decreasing the effective stress. The soil weight also decreases with the increase in pore water pressure. When downward seepage occurs, the direction of seepage force and gravity is just the same, making

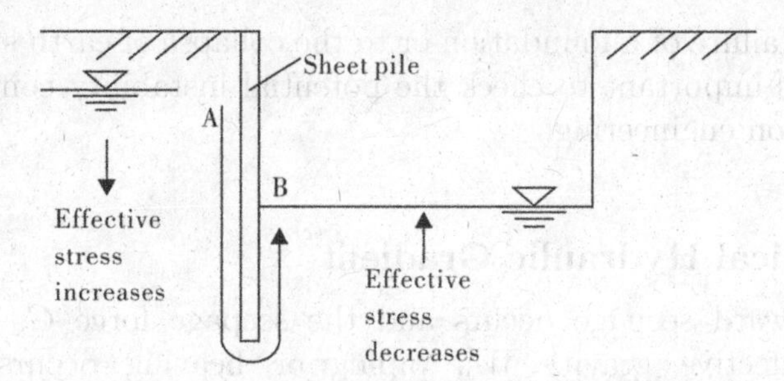


Fig. 2.10. Effects of seepage on the effective stress.

the effective stress increase. The soil particles become more compact with an increase in pore water pressure, which is an advantage for construction engineering. For example, in ancient China, there is a practice of pouring water in medium or coarse sand and applying vibration to make the sand compact. When lateral seepage occurs, it is convenient to calculate pore water pressure by using flow net.

### 2.6.2 *Upward seepage through soil*

Under water, the effective gravity per unit volume of soil mass is $W'$, that is

$$W' = \gamma_{sat} - \gamma_w = \gamma', \tag{2.24}$$

where $\gamma_{sat}$, $\gamma_w$, and $\gamma'$ are the saturated unit weight of soil, unit weight of water, and buoyant unit weight of soil, respectively (kN/m$^3$).

Consider the special case of seepage vertically upward. If the seepage force ($G_d$) exerted on the soil is equal to or greater than the effective gravity $W'(G_d = \gamma_w \cdot i \geq \gamma')$, then the soil loses its strength and behaves like a viscous fluid. The failure caused by seepage occurs. The soil state at which the strength is zero is called static liquefaction. Other names such as heaving and boiling are used to describe specific events connected to the unstable state. Boiling occurs when the upward seepage force exceeds the downward force of the silt. Heaving occurs when seepage forces push the bottom of an excavation in the upward direction. If the upward seepage forces exceed the submerged weight, the particles may be carried upward and be deposited at the ground surface and a "pipe" is formed in the soil near the surface, which is called piping. Piping refers to the subsurface "pipe-shaped erosion". Piping failure can lead to the complete failure of a foundation or to the collapse of earth structure. Thus, it is important to check the potential instability condition in construction engineering.

## 2.7 Critical Hydraulic Gradient

When upward seepage occurs and the seepage force $G_d$ is equal to the effective gravity $W'$, piping or heaving occurs in the soil mass. The value of hydraulic gradient corresponding to zero resultant body force is called the critical hydraulic gradient ($i_{\text{cr}}$). For an element of soil of volume $V$ subject to upward seepage under the critical hydraulic gradient, the seepage force (Eq. (2.22)) is therefore equal to the effective weight (Eq. (2.24)) of the element, i.e.,

$$G_d = \gamma_w V \cdot i_{\text{cr}} = \gamma' V.$$

So

$$i_{\text{cr}} = \frac{\gamma'}{\gamma_w}. \tag{2.25}$$

According to the phase relationships in Chapter 1,

$$\gamma' = (d_s - 1)(1 - n)\gamma_w = \frac{d_s - 1}{1 + e}\gamma_w.$$

Therefore,

$$i_{\text{cr}} = (d_s - 1)(1 - n) = \frac{d_s - 1}{1 + e}. \tag{2.26}$$

The symbols mentioned above have the same meaning as in Chapter 1.

The ratio $\frac{\gamma'}{\gamma_w}$ is $\approx 1.0$ for most soil. When the hydraulic gradient is $i_{\text{cr}}$, the effective normal stress on any plane will be zero, gravitational forces having been canceled out by upward seepage forces. In the case of sand, the contact forces between particles will be zero and the soil will have no strength. It should be realized that "quicksand" is not a

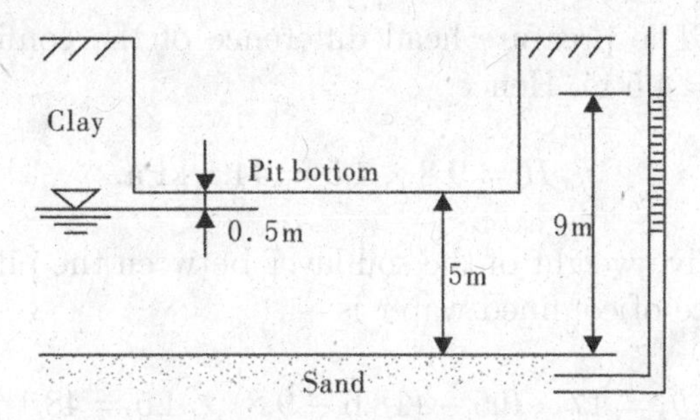


Fig. 2.11. Example 2.1.

special type of soil but simply sand through which there is an upward flow of water under a hydraulic gradient equal to or exceeding $i_{cr}$.

**Example 2.1.** As shown in Fig. 2.11, the thickness of the clay layer under the pit bottom is 5 m. Water is confined under the clay layer and the hydraulic pressure is measured by a piezometer tube. During the construction, the groundwater level is maintained at a depth of 0.5 m below the pit bottom by foundation pit dewatering. The unit weight ($\gamma$) is 17 kN/m$^3$ and the saturated unit weight ($\gamma_{sat}$) is 18.6 kN/m$^3$ of the clay. Determine whether pit bottom upheaval happens or not.

**Solution 1:** Since confined water exists, there will be upward seepage force, that is

$$G_d = \gamma_w \cdot i = \gamma_w \cdot \frac{\Delta h}{L} = 9.8 \times \frac{4.5}{4.5} = 9.8 \text{ kN/m}^3.$$

Through the 4.5-m flow path, the seepage force is expressed by surface force, that is

$$G'_d = G_d \cdot L = 9.8 \times 4.5 = 44.1 \text{ kPa}.$$

The effective weight of the soil layer above and below the groundwater is

$$W' = 17 \times 0.5 + (18.6 - 9.8) \times 4.5 = 48.1 \text{ kPa}.$$

Since $G'_d \leq W'$, pit bottom upheaval doesn't happen.

**Solution 2:** The pressure head difference of the confined water is $H = 9 - 4.5 = 4.5$ m. Hence,

$$\gamma_w H = 9.8 \times 4.5 = 44.1 \text{ kPa}.$$

The effective weight of the soil layer between the pit bottom and the top surface of confined water is

$$\sum \gamma_i h_i = 17 \times 0.5 + (18.6 - 9.8) \times 4.5 = 48.1 \text{ kPa}.$$

Since $\sum \gamma_i h_i > \gamma_w H$, pit bottom upheaval doesn't occur.

## Exercises

2.1 What are the main influences of capillary water on buildings and soils?

2.2 What is Darcy's Law and how is it used for different kinds of soils?

2.3 What is the principle of constant-head test to determine the coefficient of permeability? Why should falling-head test be used in clay soil to determine the coefficient of permeability?

2.4 What are the conditions under which heaving sand is formed? What are the main damages caused by heaving sand and how is it prevented?

2.5 In a simple constant-head permeability testing apparatus (Fig. 2.12), the sectional area of the specimen is 120 cm$^2$. The amount of water through the specimen is 150 cm$^3$ in 10 s. Determine the coefficient of permeability of the specimen.

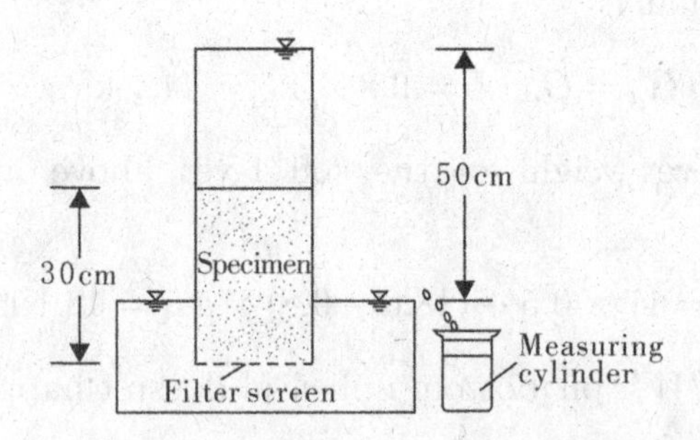


Fig. 2.12. Exercise 2.5.

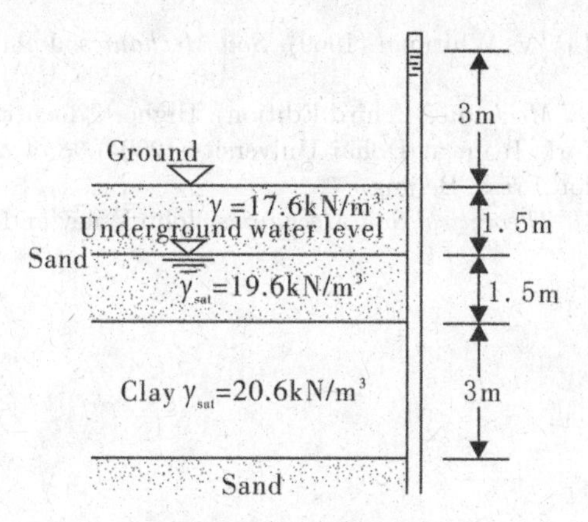


Fig. 2.13. Exercise 2.8.

2.6 A certain cohesionless soil has a void ratio $e = 0.61$ and specific gravity $d_s = 2.65$. Determine the critical hydraulic gradient when heaving occurs.

2.7 During the excavation of a foundation pit, draining water causes upward seepage. The water head difference is 60 cm, the length of water flow through the soil is 50 cm, and the saturated unit weight $\gamma_{\text{sat}} = 20.5$ kN/m. Determine whether the heaving sand happens.

2.8 One clay layer is located between two sand layers. The bulk unit weight of the sand ($\gamma$) is 17.6 kN/m$^3$, the saturated unit weight of the sand is ($\gamma_{\text{sat}}$) 19.6 kN/m$^3$, and the saturated unit weight of the clay ($\gamma_{\text{sat}}$) is 20.6 kN/m. The thickness of each soil layer is shown in Fig. 2.13. The groundwater level is at a depth of 1.5 m below the ground level. Assume that there is pressured water in the lower sand layer and the water level of the standpipe is at 3 m above the ground level. If heaving occurs at the clay layer, how much higher should the water level of the standpipe be than the ground level?

## Bibliography

J. Atkinson (1993). *An Introduction to the Mechanics of Soils and Foundations.* McGraw-Hill Book Company Europe, England.

R. F. Craig (1998). *Soil Mechanics.* E and FN Spon, London and New York.

T. W. Lambe and R. V. Whitman (1969). *Soil Mechanics.* John Wiley and Sons, New York.

H. Liao (2018). *Soil Mechanics* (Third Edition). Higher Education Press, Beijing.

Soil Mechanics Work Team at Hohai University (2004). *Soil Mechanics.* China Communication Press, Beijing.

K. Terzaghi (1943). *Theoretical Soil Mechanics.* John Wiley and Sons, New York.

# Chapter 3

# Stress Distribution in Soil

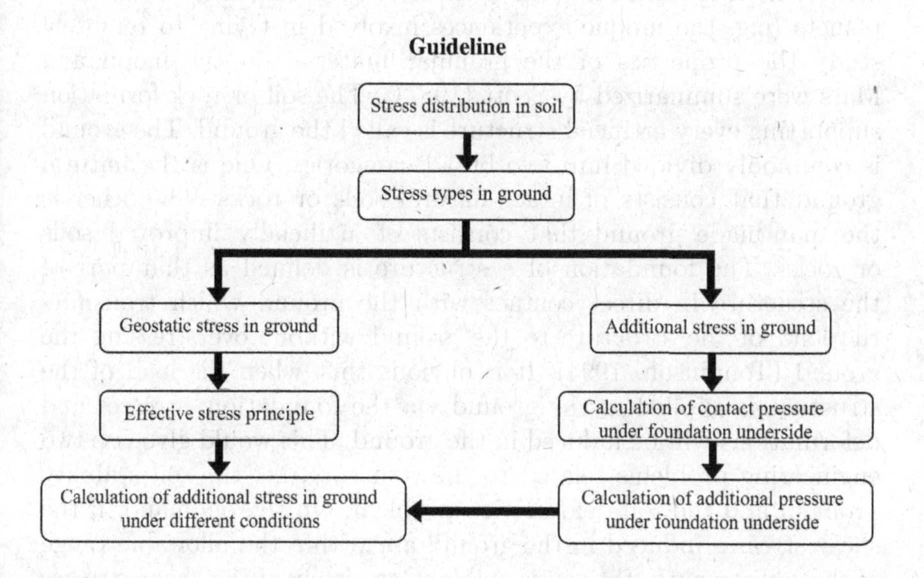

## 3.1 Introduction

### 3.1.1 *Purpose of researching the stress state of soil*

Geotechnical engineering is the application of the science of soil mechanics and rock mechanics, engineering geology, and other related disciplines to civil engineering, extractive industries (e.g. underground mining), and for the preservation and enhancement of the environment (e.g. landfills of municipal solid waste or nuclear waste) (Anon, 1999).

Geotechnical engineering is a sub-discipline within civil engineering. It covers all forms of soil-related problems. Geotechnical engineering plays a key role in all civil engineering projects, since all construction is built on or in the soil or rock on earth or on other planets (e.g. the unique experiences involved in trying to remotely study the properties of the granular material on the moon and Mars were summarized by Scott (1987)). The soil or rock formation supporting every artificial structure is called the ground. The ground is commonly divided into two broad categories. One is the natural ground that consists of intact natural soils or rocks. The other is the man-made ground that consists of artificially improved soils or rocks. The foundation of a structure is defined as that part of the structure in direct contact with the ground which transmits the load of the structure to the ground without overstressing the ground (Tomlinson, 1994). It is obvious that when the load of the structure is applied to the ground via the foundation, stresses and deformations will be induced in the ground. This would give rise two engineering problems related to the structure, i.e. the soil stability problem and the soil deformation problem. On the one hand, if the shear stresses induced in the ground are within the allowable range of the soil strength, the soil is stable. Conversely, if the shear stresses induced in a certain localized area in the ground are in excess of the allowable range of the soil strength, failure of the soil may occur, and this may result in the slippage of the global ground, leading to an overturning of the structure. On the other hand, if the deformations of the ground exceed the permissible values, then the structure may also be damaged and lose its service function, although the soil has not failed yet. Therefore, in order to ensure the safety and the normal service function of the structure, the stress distribution patterns and the deformations induced in the ground under various loading conditions must be studied.

Calculation of the ground deformations is introduced in the next chapter. This chapter presents only the calculation of the stresses in the ground soils and their distribution patterns.

Actually, the stresses in the ground soils can be divided into two categories in terms of their origin as follows:

- One is the *effective overburden pressure*, i.e. the stresses due to the effective self-weight of the soil. It is also known as the geostatic pressure or the at-rest *in situ* stress. Generally speaking, compression of the soils caused by their effective self-weight has been already happening over a very long period of time; thus, soil deformation due to the effective overburden pressure would not be induced further. However, it is an exception for newly deposited sedimentation or reclaimed soils such as recent hydraulic fills.
- The other is the *stress increment*, i.e. the stress in the interior of the ground due to exterior (static or dynamic) load. It is also known as additional stress. Stress increment may be the major cause of soil instability and ground deformations. The magnitude of the stress increment is dependent not only on the location of the calculation point but also on the magnitude and distribution pattern of the contact pressure between the foundation and the ground.

This chapter first presents the calculation of the effective overburden pressure, followed by the distribution pattern and the corresponding calculation method of the contact pressure beneath the foundation. Finally, the calculation of the stress increment due to various loading conditions is introduced.

### 3.1.2 *Stress state of soil*

#### 3.1.2.1 *The basic assumption of the stress calculation*

**(1) Continuous medium assumption:** Elastic theory requires that soil be a continuous medium, but soil comprises three phases of matter.

**(2) Linear elastic body assumption:** The stress–strain curve of soil is linear when the stress does not exceed the yield point.

(3) **Homogeneous and isotropic body:** When the variations in the property of soil layer are not too large, soil is a homogeneous and isotropic body.

We can adopt the solutions in elastic mechanics to solve the stress in elastic soil and the analysis method could be the most simple and easy method by which to draw a graph.

3.1.2.2 *The common stress state in ground*

**(1) The general state of stress — Three-dimensional problem**

The stress state of groundwork is a three-dimensional problem under load acting, and an elemental soil mass with sides measuring $dx$, $dy$, and $dz$ is shown in Fig. 3.1. Parameters $\sigma_x$, $\sigma_y$, and $\sigma_z$ are the normal stresses acting on the planes normal to the $x$, $y$, and $z$ axes, respectively. The normal stresses are considered positive when they are directed onto the surface. Parameters $\tau_{xy}$, $\tau_{yx}$, $\tau_{yz}$, $\tau_{zy}$, $\tau_{xz}$, and $\tau_{zx}$ are shear stresses. All shear stresses are positive in Fig. 3.1.

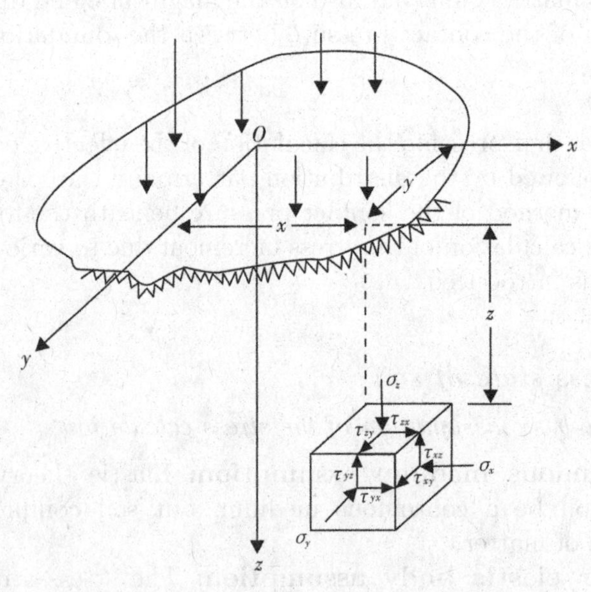


Fig. 3.1. The general stress state of soil.

The stress at any point has nine stress components and can be written in matrix form as follows:

$$\sigma_{ij} = \begin{bmatrix} \sigma_x & \tau_{xy} & \tau_{xz} \\ \tau_{yx} & \sigma_y & \tau_{yz} \\ \tau_{zx} & \tau_{zy} & \sigma_z \end{bmatrix}. \tag{3.1}$$

For equilibrium, $\tau_{xy} = \tau_{yx}$, $\tau_{yz} = \tau_{zy}$, $\tau_{xz} = \tau_{zx}$, the elemental soil mass has six stress components, i.e. $\sigma_x$, $\sigma_y$, $\sigma_z$, $\tau_{xy}$, $\tau_{xz}$, and $\tau_{yz}$.

**(2) Plane strain conditions — Two-dimensional problem**

Assume there is a strip foundation with infinite length, and the ratio of the length $l$ to the width $b$ of the foundation is given as $l/b \geq 10$. In this case, it can be considered as a plane strain problem and any cross-section (perpendicular to the long axis) of the foundation is in the same stress state. The soils with plane strain conditions will appear deformation in the plane of $x$ and $z$, but no deformation is seen in the direction of $y$, i.e. $\varepsilon_y = 0$ and $\tau_{yx} = \tau_{yz} = 0$. The unit body has five stress components with symmetrical characteristics, i.e. $\sigma_{xx}$, $\sigma_{yy}$, $\sigma_{zz}$, $\tau_{xz}$, $\tau_{zx}$, and is written in matrix form as follows:

$$\sigma_{ij} = \begin{bmatrix} \sigma_{xx} & 0 & \tau_{xz} \\ 0 & \sigma_{yy} & 0 \\ \tau_{zx} & 0 & \sigma_{zz} \end{bmatrix}. \tag{3.2}$$

**(3) State of the confining stress — One-dimensional problem**

The horizontal ground is a semi-infinite half-space body and the geostatic stress in semi-infinite elastic ground is only related to $z$, namely, soil element without lateral motion and this is defined as the state of the confining stress. Thus, the random vertical plane is a symmetrical plane. In this case, $\tau_{xy} = \tau_{yz} = \tau_{zx} = 0$, and the stress matrix is given as follows:

$$\sigma_{ij} = \begin{bmatrix} \sigma_{xx} & 0 & 0 \\ 0 & \sigma_{yy} & 0 \\ 0 & 0 & \sigma_{zz} \end{bmatrix}, \tag{3.3}$$

where $\sigma_{xx} = \sigma_{yy}$ due to $\varepsilon_x = \varepsilon_y = 0$, and it is a direct ratio with respect to $\sigma_z$.

## 3.2 Stresses Due to Self-Weight

When soils are exposed to loading, stress and deformation are produced. Generally, data concerning internal stress conditions are used to determine deformation, i.e. the deformation can be determined by using stress. Stresses within soil are caused by the external loads applied to the soil as well as the self-weight of the soil. The pattern of stresses caused by the applied loads is usually quite complex.

### 3.2.1 *Vertical geostatic stress*

In the case just described, there are no shear stresses upon vertical and horizontal planes within the soil. Hence, the vertical geostatic stress at any depth can be computed simply by considering the weight of soil above that depth.

Thus, if the unit weight of the soil is constant with depth, we have

$$\sigma_z = \gamma_1 h, \tag{3.4}$$

where $h$ is the depth and $\gamma$ is the total unit weight of the soil. In this case, the vertical stress will vary linearly with depth, as shown in Fig. 3.2(a). The distribution of vertical stress along the depth shows a triangle shape. If the soil is stratified and the unit weight is different for each stratum, then the vertical stress can conveniently be computed by means of the summation, as shown in

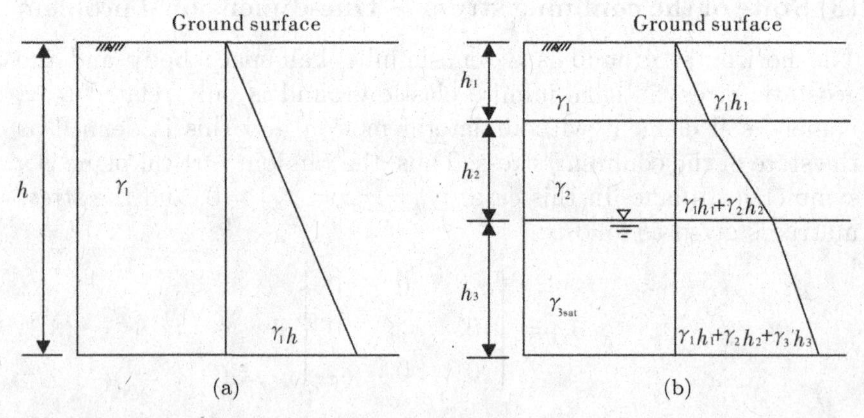


Fig. 3.2. Geostatic stress in soil. (a) Geological section and (b) geostatic stress distribution.

Fig. 3.2(b), i.e.

$$\sigma_z = \sum_{i=1}^{n} \gamma_i h_i. \tag{3.5}$$

If the unit weight of the soil varies continuously with depth, then the vertical stress can be evaluated by means of the integral.

**Example 3.1.** A foundation is composed of multiple layers of soils, which is shown in the geological section, Fig. 3.3(a). Calculate and draw geostatic stress distribution of the soils along the depth direction.

**Solution.** Consider five points A, B, C, D, and E as the reference points to calculate the geostatic stress, as shown in Fig. 3.3(b).

$$\begin{aligned}
\text{A}: &\ \sigma_{czA} = 0, \\
\text{B}: &\ \sigma_{czB} = \gamma_B h_B = 19 \times 3 = 57.0\,\text{kPa}, \\
\text{C}: &\ \sigma_{czC} = \gamma_B h_B + \gamma'_C h_C = 19 \times 3 \\
&\quad + (20.5 - 10) \times 2.2 = 80.1\,\text{kPa}, \\
\text{D (up)}: &\ \sigma_{czD\text{up}} = \gamma_B h_B + \gamma'_C h_C + \gamma'_{D\text{up}} h_D = 80.1 \\
&\quad + (19.2 - 10) \times 2.5 = 103.1\,\text{kPa},
\end{aligned}$$

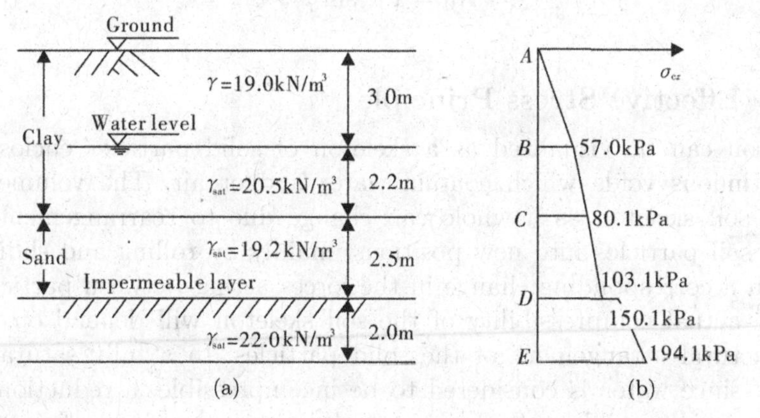


Fig. 3.3. Example 3.1. (a) Geological section and (b) geostatic stress distribution.

$$\text{D(down)}: \sigma_{cz\text{Ddown}} = \gamma_B h_B + \gamma_{\text{sat}\,C} h_C + \gamma_{\text{sat}D\text{down}} h_D$$
$$= 19 \times 3 + 20.5 \times 2.5 + 19.2 \times 2.5 = 150.1\,\text{kPa},$$
$$\text{E}: \sigma_{czE} = \gamma_B h_B + \gamma_{\text{sat}\,C} h_C + \gamma_{\text{sat}D\text{down}} h_D + \gamma_{\text{sat}E} h_E$$
$$= 150.1 + 22 \times 2 = 194.1\,\text{kPa}.$$

### 3.2.2 *Horizontal geostatic stress*

The ratio of horizontal to vertical stress for a mass of soil in a state of rest is expressed by a factor called the coefficient of earth pressure at rest or lateral stress ratio at rest and is denoted by the symbol $K_0$, i.e.

$$K_0 = \frac{\sigma_{h0}}{\sigma_{v0}}. \tag{3.6}$$

The value of $K_0$ can be determined experimentally by means of a triaxial test in which the axial stress and the all-round pressure are increased simultaneously such that the lateral strain in the specimen is maintained at zero (the hydraulic triaxial apparatus is most suitable for this purpose).

For normally consolidated soils, the value of $K_0$ can be related approximately to the effective stress parameter $\varphi'$ by the following formula proposed by Jaky:

$$K_0 = 1 - \sin\varphi'. \tag{3.7}$$

## 3.3 Effective Stress Principle

A soil can be visualized as a skeleton of solid particles enclosing continuous voids which contain water and/or air. The volume of the soil skeleton as a whole can change due to rearrangement of the soil particles into new positions, mainly by rolling and sliding, with a corresponding change in the forces acting between particles. The actual compressibility of the soil skeleton will depend on the structural arrangement of the solid particles. In a fully saturated soil, since water is considered to be incompressible, a reduction in volume is possible only if some of the water can escape from the voids. In dry or a partially saturated soil, a reduction in volume is always possible due to compression of the air in the voids, provided there is scope for particle rearrangement.

Shear stress can be resisted only by the skeleton of solid particles, by means of forces developed at the interparticle contacts. Normal stress may be resisted by the soil skeleton through an increase in the interparticle forces. If the soil is fully saturated, the water filling the voids can also withstand normal stress by an increase in pressure.

The importance of the forces transmitted through the soil skeleton from particle to particle was recognized in 1923 when Terzaghi presented the principle of effective stress, an intuitive relationship based on experimental data. The principle applies only to fully saturated soils and relates the following three stresses:

(1) The effective normal stress ($\sigma'$) on the plane, representing the stress transmitted through the soil skeleton only. It is the total particle interaction force per unit area. It controls (a) deformation and (b) shear strength of soil mass.
(2) The total normal stress ($\sigma$) on a plane within the soil mass, being the total (particle + water) force per unit area.
(3) The pore water pressure ($u$), being the pressure of the water filling the void space between the solid particles.

The relationship between the three stresses is given as follows:

$$\sigma' = \sigma - u. \tag{3.8}$$

The principle can be represented by the following physical model. Consider a "plane a–a" in a fully saturated soil, passing through points of interparticle contact only, as shown in Fig. 3.4. Then, the effective normal stress is interpreted as the sum of all the components $N'$ within the area $A$, divided by the area $A$, i.e.

$$\sigma' = \frac{\sum N'}{A}. \tag{3.9}$$

The total normal stress is given by

$$\sigma = \frac{P}{A}. \tag{3.10}$$

If a point contact is assumed between the particles, the pore water pressure will act on the plane over the entire area $A$. Then, for

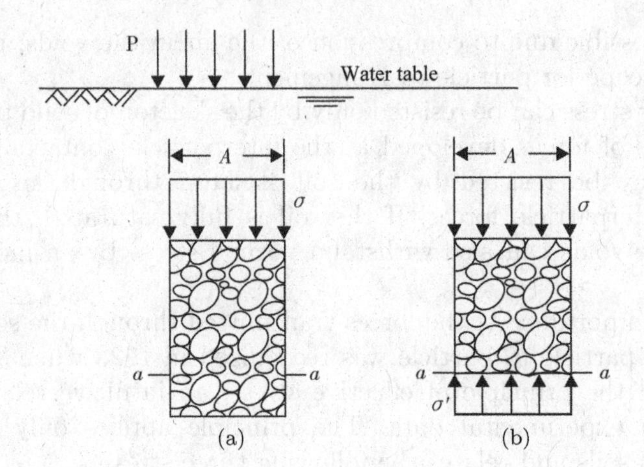


Fig. 3.4. Interpretation of effective stress.

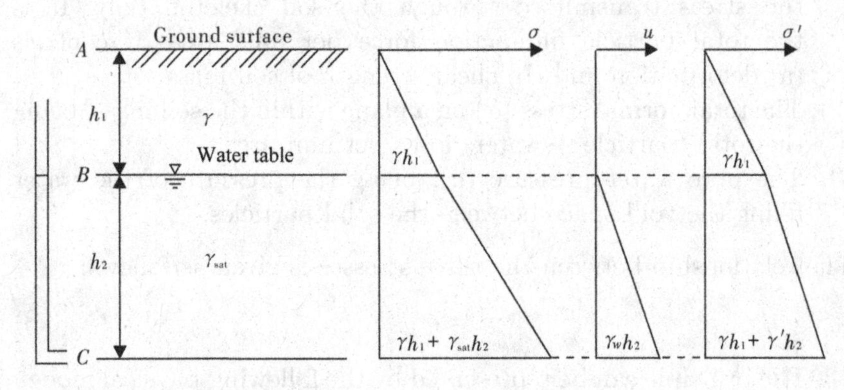


Fig. 3.5. Effective vertical stress due to self-weight of soil.

equilibrium in the direction normal to a–a, we have

$$P = \sum N' + uA \quad \text{or} \quad \frac{P}{A} = \frac{\sum N'}{A} + u, \text{ i.e. } \sigma = \sigma' + u. \tag{3.11}$$

Consider a soil mass having a horizontal surface and with the water table at the surface level, as shown in Fig. 3.5. The total vertical stress (i.e. the total normal stress on a horizontal plane) at depth $C$ is equal to the weight of all the materials (solids + water) per unit area above that depth, i.e.

$$\sigma_v = \gamma h_1 + \gamma_{\text{sat}} h_2. \tag{3.12}$$

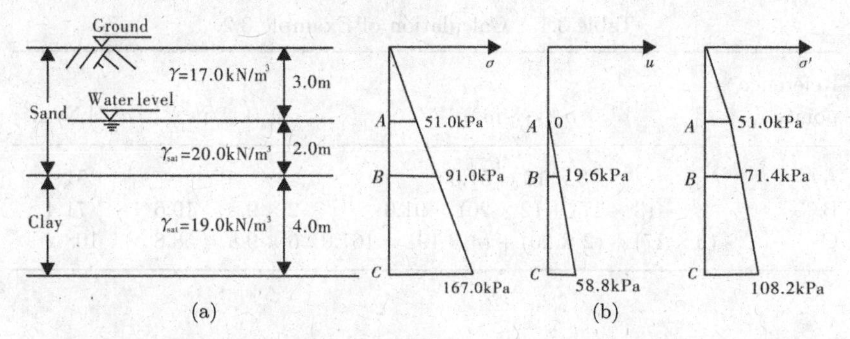


Fig. 3.6. Example 3.2. (a) Geological section and (b) $\sigma$, $u$, $\sigma'$ distribution.

The pore water pressure at any depth will be hydrostatic since the void space between the solid particles is continuous, so at depth $C$, we have

$$u = \gamma_w h_2. \tag{3.13}$$

Hence, from the Terzaghi's effective stress principle, the effective vertical stress at depth $C$ will be

$$\sigma_v' = \sigma_v - u = \gamma h_1 + (\gamma_{\text{sat}} - \gamma_w) h_2 = \gamma h_1 + \gamma' h_2, \tag{3.14}$$

where $\gamma'$ is the buoyant unit weight of the soil.

**Example 3.2.** A foundation is composed of multiple layers of soils, which is shown in the geological section, Fig. 3.6(a). Calculate and draw the total normal stress distribution, pore water pressure distribution, and the effective normal stress distribution of the soils along the depth direction.

**Solution.** Take three points A, B, and C as the reference points to calculate $\sigma$, $u$, and $\sigma'$, as shown in Table 3.1. The distribution of the total normal stress, pore water pressure, and effective normal stress of the soils along the depth direction are shown in Fig. 3.6(b).

## 3.4 Contact Pressure between the Foundation and the Ground

### 3.4.1 *Contact pressure distribution*

Virtually every structure is supported by ground soils or rocks. The major function of the foundation of structure is to transmit the load of the structure to the supporting ground. Therefore, the analysis of

Table 3.1. Calculation of Example 3.2.

| Reference points | $\sigma$ (kN/m$^2$) | $u$ (kN/m$^2$) | $\sigma'$ (kN/m$^2$) |
|---|---|---|---|
| A | $3 \times 17 = 51.0$ | 0 | 51 |
| B | $(3 \times 17) + (2 \times 20) = 91.0$ | $2 \times 9.8 = 19.6$ | 71.4 |
| C | $(3 \times 17) + (2 \times 20) + (4 \times 19) = 167.0$ | $6 \times 9.8 = 58.8$ | 108.2 |

the interaction between a structural foundation and the supporting ground soil media is of primary importance to both structural and geotechnical engineering.

Contact pressure is the intensity of loading transmitted from the underside of a foundation to the ground soil (Whitlow, 2001). The magnitude and the distribution pattern of the contact pressure have an important impact on the stress increment induced in the ground. The magnitude and the distribution pattern of the contact pressure depend on many factors such as the magnitude and distribution of the structure load applied, the rigidity and embankment depth of the foundation, and the soil properties.

It has been found from tests that for a foundation with a very low rigidity or for a flexible foundation, the magnitude and the distribution pattern of the contact pressure are the same as those of the load applied on the foundation. This is because the foundation is compatible to the deformation of the ground soil. When the load on the foundation is uniformly distributed, the contact pressure (normally denoted as the reaction force on the underside of a foundation, ditto) is also uniformly distributed, as shown in Fig. 3.7(a). When the load distribution is trapezoidal, the contact pressure distribution is also trapezoidal, as shown in Fig. 3.7(b).

For a rigid foundation that cannot be compatible to the ground deformation due to a significant difference in rigidity, the distribution of the contact pressure varies with the magnitude of the applied load, the embedment depth of the foundation, and the properties of the ground soil. For instance, when a centric load is applied to the rigid strip foundation founded on the surface of a sandy ground, the contact pressure at the centerline of the foundation is maximum, the contact pressure at the edge of the foundation is zero, and its

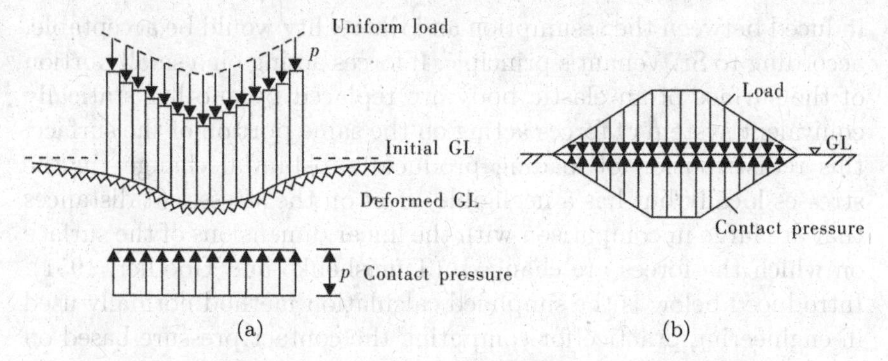


Fig. 3.7. Contact pressure distribution beneath a flexible foundation.

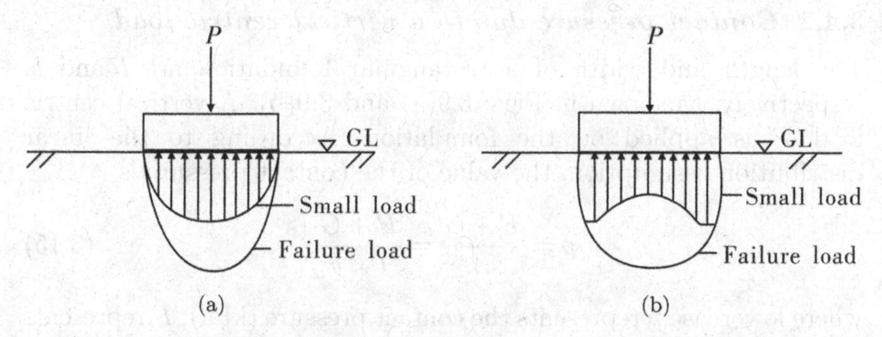


Fig. 3.8. Contact pressure distribution beneath a rigid foundation.

distribution looks like a parabolic curve, as shown in Fig. 3.8(a). This is because no cohesion is available among sand particles. When a centric load is applied to the rigid strip foundation founded on the surface of a clayey ground, some loads can be carried on the edge of the foundation due to cohesion of the clayey soils. Therefore, when the applied load is relatively small, high contact pressure will be imposed on the edge of the foundation and low contact pressure will be imposed on the center of the foundation. The distribution curve is shaped like a saddle. When the load is increased gradually to failure load, the distribution curve of the contact pressure becomes higher at the center and lower at the edge of the foundation and is shaped similar to a bowl, as shown in Fig. 3.8(b).

Empirically, when the width of the rigid foundation is small and the applied load is relatively small, the contact pressure distribution follows approximately a linear distribution assumption. The error induced between the assumption and the reality would be acceptable, according to St. Venant's principle. If forces acting on a small portion of the surface of an elastic body are replaced by another statically equivalent system of forces acting on the same portion of the surface, this redistribution of loading produces substantial changes in the stresses locally but has a negligible effect on the stresses at distances that are large in comparison with the linear dimensions of the surface on which the forces are changed. (Timoshenko and Goodier, 1951). Introduced below is the simplified calculation method normally used in engineering practice for computing the contact pressure based on the linear distribution assumption.

### 3.4.2 *Contact pressure due to a vertical centric load*

The length and width of a rectangular foundation are $l$ and $b$, respectively, as shown in Figs. 3.9(a) and 3.9(b). A vertical centric load $P$ is applied on the foundation. According to the linear distribution assumption, the value of the contact pressure is

$$p = \frac{F+G}{A} = \frac{F+G}{l \times b}, \tag{3.15}$$

where lowercase $p$ represents the contact pressure (kPa); $F$ represents the vertical load on the upside of the foundation (kN); $G$ represents the self-weight of the foundation and the soil weight on the steps of the foundation, and generally, the value $20\,\text{kN/m}^3$ is adopted as the average unit weight; and $A = l \times b$ represents the area of the foundation ($\text{m}^2$). $l$ and $b$ represent the length and width of the foundation.

If the foundation is oblong (theoretically, when $l/b$ approaches infinity, it is called a strip foundation; practically, when $l/b$ is greater than or equal to 10, it can be considered as a strip foundation), a free body of 1 m unit length can be truncated in the longitudinal direction of the foundation for the calculation, as shown in Fig. 3.9(c). In this scenario, the contact pressure is given as

$$p = \frac{F+G}{b}, \tag{3.16}$$

where $b$ represents the width of the foundation (m); the other symbols have the same meaning as presented before.

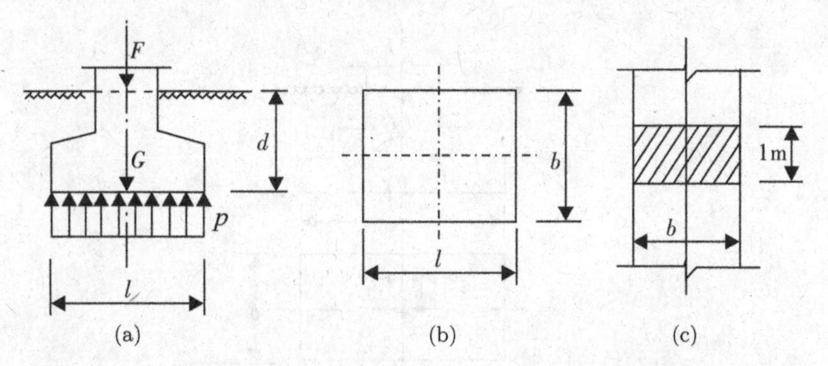


Fig. 3.9. Contact pressure distribution due to a vertical centric load.

### 3.4.3 *Contact pressure due to a one-way vertical eccentric load*

When a one-way eccentric load is applied to a rectangular foundation (as shown in Fig. 3.10), the contact pressure at any arbitrary point can be calculated using the formula of eccentric compression in mechanics of materials, as given by

$$p_{\substack{\max \\ \min}} = \frac{F+G}{A} \pm \frac{M}{W} = \frac{F+G}{A}\left(1 \pm \frac{6e}{l}\right), \qquad (3.17)$$

where $p_{\max}$ and $p_{\min}$ represent the maximum and minimum contact pressures on both sides of the underside of the foundation (kPa); $M$ represents the moment of the eccentric load about the $Y - Y$ axis; $W$ represents the resisting moment of the underside of foundation, and $W = \frac{bl^2}{6}$ if the area is rectangular (m$^3$); and $e$ is the offsetting of the eccentric load line to the $Y$–$Y$ axis.

It can be seen from Eq. (3.17) that when the resultant offsetting $e$ is less than $l/b$, the distribution curve of the contact pressure is trapezoidal. When the resultant offsetting $e$ is equal to 1/6, $p_{\min}$ is zero and the distribution curve of the contact pressure is triangular shape. When the resultant offsetting $e$ is greater than 1/6, $p_{\min}$ is less than zero and tension force would appear on one side of the underside of the foundation, as shown in Figs. 3.10(a)–3.10(c). Generally speaking, the tension force on the underside of the foundation is not allowed in the engineering practice; therefore, when designing the size of the foundation, the resultant offsetting should satisfy a criterion of $e$ less than 1/6, for the sake of safety. Because

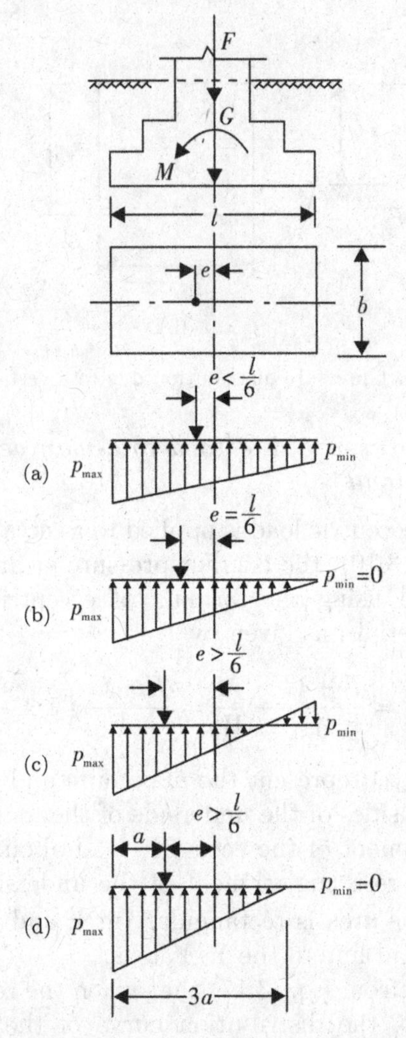


Fig. 3.10. Contact pressure distribution due to a one-way vertical eccentric load.

soil cannot bear tension force, the contact pressure is adjusted. The calculation principle is that the composite pressure of foundation base pressure is identical to the total load (as shown in Fig. 3.10(d)), and the formula of the maximum contact pressure $p_{max}$ is given by

$$p_{\max} = \frac{2(F+G)}{3ba}, \tag{3.18}$$

where $a$ represents the distance between the action point of eccentric load and the edge of the $p_{\max}$, $a = \frac{l}{2} - e$ (m).

Similarly, for a strip foundation, the maximum and minimum contact pressures of the underside of the foundation are given as

$$p_{\substack{\max \\ \min}} = \frac{F+G}{A}\left(1 \pm \frac{6e}{b}\right). \quad (3.19)$$

### 3.4.4 *Contact pressure due to a two-way vertical eccentric load*

When a two-way eccentric load is applied to a rectangular foundation (as shown in Fig. 3.11), the contact pressure at any arbitrary point can be calculated using the formula of eccentric compression in mechanics of materials, as given by

$$p_{\substack{\max \\ \min}} = \frac{F+G}{A} \pm \frac{M_x y}{I_x} \pm \frac{M_y x}{I_y}, \quad (3.20)$$

where $M_x = (F+G)e_y$ represents the moment of the eccentric load about the $X$–$X$ axis ($e_y$ is the offsetting of the eccentric load line to the $X$–$X$ axis); $M_y = (F+G)e_x$ represents the moment of the eccentric load about the $Y$–$Y$ axis ($e_x$ is the offsetting of the eccentric load line to the $Y$–$Y$ axis); $I_x = bl^3/12$ represents the moment of inertia of the area of the underside of the foundation about the $X$–$X$ axis; $I_y = lb^3/12$ represents the moment of inertia of the area of the underside of the foundation about the $Y$–$Y$ axis.

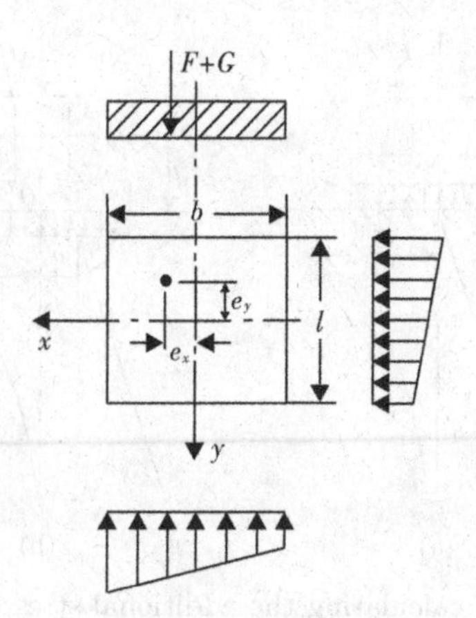


Fig. 3.11. Contact pressure distribution due to a two-way vertical eccentric load.

### 3.4.5 *Additional stress on the underside of the foundation*

Additional stress is defined as the increased pressure in the foundation due to building the architecture, as shown in Fig. 3.12.

(1) When the foundation is constructed above the ground surface (Fig. 3.12(a)), the additional stress on the underside of the foundation $p_0$ is the contact pressure of the underside of the foundation $p$, that is

$$p_0 = p. \tag{3.21}$$

(2) When the foundation is constructed at some depth under the ground surface (Fig. 3.12(b)), the additional stress on the underside of the foundation $p_0$ is calculated by the following equation:

$$p_0 = p - \sigma_c = p - \gamma_0 d, \tag{3.22}$$

where $p$ is the contact pressure of the underside of the foundation (kPa), $\sigma_c$ is the overburden pressure at the foundation base (kPa), $d$ is the depth from the ground surface to the underside of the foundation (m), and $\gamma_0$ is the weighted average unit weight of the soil layers above the foundation base (kPa) and is given as follows:

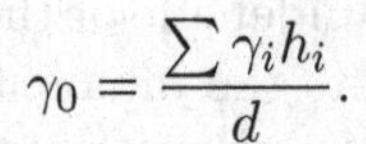

$$\gamma_0 = \frac{\sum \gamma_i h_i}{d}.$$

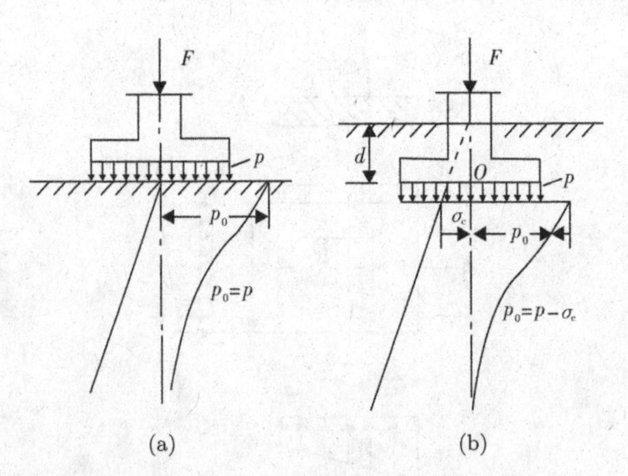


Fig. 3.12. Diagram for calculating the additional stress on the underside of the foundation.

The increase in pressure is triggered by constructing the structure after the earth is excavated, so it is the contact pressure of the underside of the foundation after subtracting the original overburden pressure.

## 3.5 Additional Stress in Ground Base

Currently, in order to obtain the stress increment on the ground due to additional load, the ground soil is generally assumed to be a continuous, homogeneous, isotropic, and fully elastic material. The stress increment can then be calculated using the basic formulae of the elasticity theory. In addition, the stress increment problem can be divided into spatial (three-dimensional) problems and plane (two-dimensional) problems in terms of their nature. If the stress is a function of the three stress components $(\sigma_x, \sigma_y, \sigma_z)$, the stress increment problem is a spatial problem. For example, the calculation of the stress increment on the undersides of rectangular and circular foundations, etc., is a spatial problem. If stress is a function of the two stress components ($\sigma_x$, $\sigma_y$ or $\sigma_x$, $\sigma_z$ or $\sigma_y$, $\sigma_z$), the stress increment problem is a plane problem. The calculation of the stress increment on the underside of a strip foundation is classified under this category. The foundations of most of the water reservoir engineering constructions such as dams and retaining walls, etc., are also strip foundations.

### 3.5.1 *Stress increment due to a vertical concentrated (point) load*

When a vertical concentrated load $F$ is applied on the surface of an elastic half-space as shown in Fig. 3.13, the six stress components $\sigma_x$, $\sigma_y$, $\sigma_z$, $\tau_{xy} = \tau_{yx}$, $\tau_{yz} = \tau_{zy}$, $\tau_{xz} = \tau_{zx}$ at an arbitrary point $M$ in the interior of the elastic body can estimated by using the elasticity theory, as given in the following equations:

$$\sigma_z = \frac{3F}{2\pi} \cdot \frac{z^3}{R^5}, \tag{3.23a}$$

$$\sigma_y = \frac{3F}{2\pi} \cdot \left\{ \frac{y^2 z}{R^5} + \frac{1-2v}{3} \left[ \frac{1}{R(R+Z)} - \frac{(2R+z)y^2}{(R+z)^2 R^3} - \frac{z}{R^3} \right] \right\}, \tag{3.23b}$$

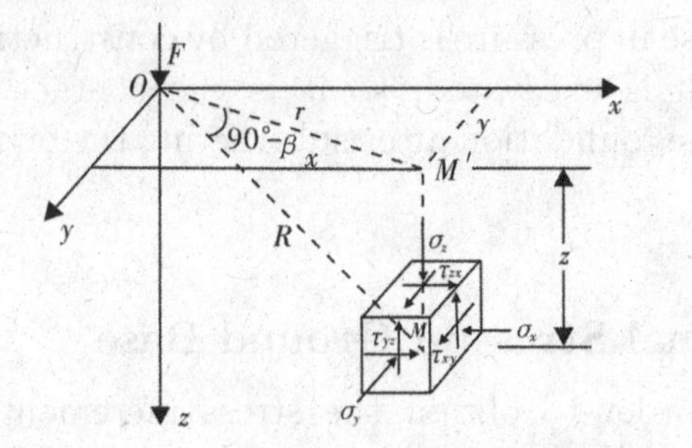


Fig. 3.13. Stress state of the soil due to a vertical concentrated load.

$$\sigma_x = \frac{3F}{2\pi} \cdot \left\{ \frac{x^2 z}{R^5} + \frac{1-2v}{3} \left[ \frac{1}{R(R+z)} - \frac{(2R+z)x^2}{(R+z)^2 R^3} - \frac{z}{R^3} \right] \right\}, \tag{3.23c}$$

$$\tau_{xy} = \frac{3F}{2\pi} \cdot \left[ \frac{xyz}{R^5} + \frac{1-2v}{3} \cdot \frac{(2R+z)xy}{(R+z)^2 R^3} \right], \tag{3.23d}$$

$$\tau_{zy} = \frac{3F}{2\pi} \cdot \frac{yz^2}{R^5}, \tag{3.23e}$$

$$\tau_{zx} = \frac{3F}{2\pi} \cdot \frac{xz^2}{R^5}. \tag{3.23f}$$

The above equations are the well-known Boussinesq solutions. They are basic formulae for solving the stress increment in the ground. In soil mechanics (as against in elasticity), a uniform sign convention has been used in that the following are considered as positive: compressive stress, reduction in length or volume, and displacement in the positive coordinate direction.

In soil mechanics, the vertical (or normal) stress component $\sigma_z$ on the horizontal plane is of special importance, as it is the major cause of the compressive deformation of the ground soil. Therefore, the calculation of the additional stress and the analyses of its distribution pattern are discussed below.

In the light of the geometrical relationship $R^2 = r^2 + z^2$ in Fig. 3.13, Eq. (3.23a) can be rewritten as given in the following form:

$$\sigma_z = \frac{3F}{2\pi} \cdot \frac{z^3}{R^5} = \frac{3F}{2\pi \cdot z^2} \cdot \frac{1}{\left[1 + \left(\frac{r}{z}\right)^2\right]^{\frac{5}{2}}} = \alpha \cdot \frac{F}{z^2}, \tag{3.24}$$

where

$$\alpha = \frac{3}{2\pi} \cdot \frac{1}{\left[1 + \left(\frac{r}{z}\right)^2\right]^{\frac{5}{2}}}$$

is the coefficient of the additional stress beneath the underside of the foundation due to a vertical concentrated load. It is a function of $r/z$ and it can be read off in Table 3.2.

It can be seen from Eq. (3.24) that the following three conclusions can be obtained:

(1) On the concentrated load line ($r = 0, \alpha = \frac{3}{2\pi}, \sigma_z = \frac{3}{2\pi} \cdot \frac{p}{z^2}$), the additional stress decreases with increasing depth $z$, as shown in Fig. 3.14.
(2) At a certain distance $r$ away from the concentrated load line, the additional stress $\sigma_z$ is zero at the ground surface, and it increases gradually with increasing depth. However, $\sigma_z$ decreases with increasing depth, as shown in Fig. 3.14.
(3) On a horizontal plane at a certain depth $z$, the additional stress decreases with increasing $r$, as shown in Fig. 3.14.

### 3.5.2 *Additional stress beneath the corners of the underside of a rectangular foundation due to a vertical uniform load*

When a vertical uniform load (hereby referred to as the compressive stress, ditto) is applied to the underside of a rectangular foundation, the additional stress under the corners of the foundation can be calculated by integrating the basic Eq. (3.24) with respect to the whole rectangular area, as shown in Fig. 3.15. If the vertical uniform load intensity on the underside of a foundation is $p$, then the acting force d$p$ on the infinitesimal area d$x$d$y$ is $p$d$x$d$y$ and it can be

Table 3.2. The coefficient of the additional stress $\alpha$ due to a vertical concentrated load.

| $r/z$ | $\alpha$ | $r/z$ | $\alpha$ | $r/z$ | $\alpha$ | $r/z$ | $\alpha$ | $r/z$ | $\alpha$ | $r/z$ | $\alpha$ | $r/z$ | $\alpha$ |
|---|---|---|---|---|---|---|---|---|---|---|---|---|---|
| 0.00 | 0.4775 | 0.30 | 0.3849 | 0.60 | 0.2214 | 0.90 | 0.1083 | 1.20 | 0.0513 | 1.50 | 0.0251 | 2.00 | 0.0085 |
| 0.02 | 0.4770 | 0.32 | 0.3742 | 0.62 | 0.2117 | 0.92 | 0.1031 | 1.22 | 0.0489 | 1.54 | 0.0225 | 2.10 | 0.0070 |
| 0.04 | 0.4756 | 0.34 | 0.3632 | 0.64 | 0.2024 | 0.94 | 0.0981 | 1.24 | 0.0466 | 1.58 | 0.0209 | 2.20 | 0.0058 |
| 0.06 | 0.4732 | 0.36 | 0.3521 | 0.66 | 0.1934 | 0.96 | 0.0933 | 1.26 | 0.0443 | 1.60 | 0.0200 | 2.40 | 0.0040 |
| 0.08 | 0.4699 | 0.38 | 0.3408 | 0.68 | 0.1846 | 0.98 | 0.0887 | 1.28 | 0.0422 | 1.64 | 0.0183 | 2.60 | 0.0029 |
| 0.10 | 0.4657 | 0.40 | 0.3294 | 0.70 | 0.1762 | 1.00 | 0.0844 | 1.30 | 0.0402 | 1.68 | 0.0167 | 2.80 | 0.0021 |
| 0.12 | 0.4607 | 0.42 | 0.3181 | 0.72 | 0.1681 | 1.02 | 0.0803 | 1.32 | 0.0384 | 1.70 | 0.0160 | 3.00 | 0.0015 |
| 0.14 | 0.4548 | 0.44 | 0.3068 | 0.74 | 0.1603 | 1.04 | 0.0764 | 1.34 | 0.0365 | 1.74 | 0.0147 | 3.50 | 0.0007 |
| 0.16 | 0.4482 | 0.46 | 0.2955 | 0.76 | 0.1527 | 1.06 | 0.0727 | 1.36 | 0.0348 | 1.78 | 0.0135 | 4.00 | 0.0004 |
| 0.18 | 0.4409 | 0.48 | 0.2843 | 0.78 | 0.1455 | 1.08 | 0.0691 | 1.38 | 0.0332 | 1.80 | 0.0129 | 4.50 | 0.0002 |
| 0.20 | 0.4329 | 0.50 | 0.2733 | 0.80 | 0.1386 | 1.10 | 0.0658 | 1.40 | 0.0317 | 1.84 | 0.0119 | 5.00 | 0.0001 |
| 0.22 | 0.4242 | 0.52 | 0.2625 | 0.82 | 0.1320 | 1.12 | 0.0628 | 1.42 | 0.0302 | 1.88 | 0.0109 | | |
| 0.24 | 0.4151 | 0.54 | 0.2518 | 0.84 | 0.1257 | 1.14 | 0.0595 | 1.44 | 0.0288 | 1.90 | 0.0106 | | |
| 0.26 | 0.4054 | 0.56 | 0.2414 | 0.86 | 0.1196 | 1.16 | 0.0567 | 1.46 | 0.0275 | 1.94 | 0.0097 | | |
| 0.28 | 0.3954 | 0.58 | 0.2313 | 0.88 | 0.1138 | 1.18 | 0.0539 | 1.48 | 0.0263 | 1.98 | 0.0089 | | |

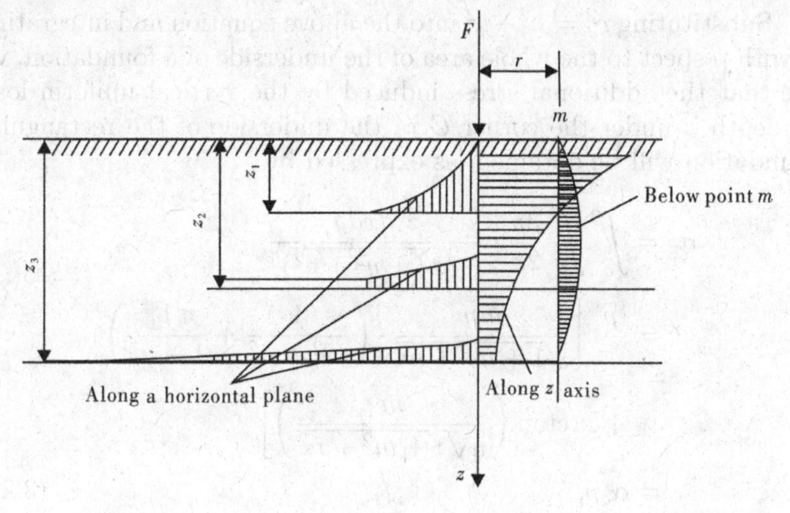


Fig. 3.14. Distribution of stress increment due to a point load.

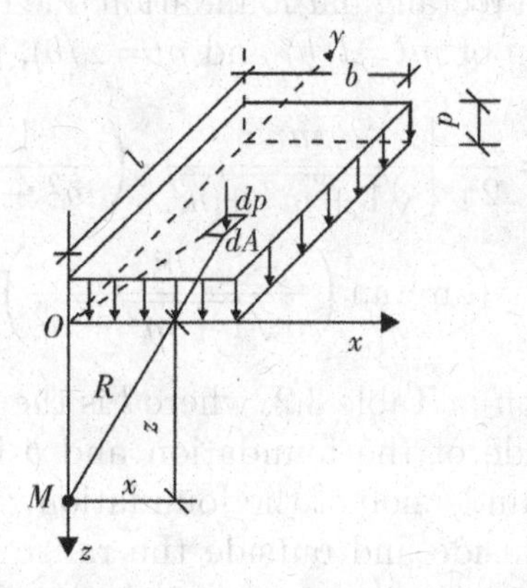


Fig. 3.15. Underside of a rectangular foundation subjected to vertical uniform load.

considered as a concentrated point load. Therefore, the additional stress at a depth $z$ under the foundation corner $O$ induced by this point load is given by

$$d\sigma_z = \frac{3p}{2\pi} \cdot \frac{1}{\left[1 + \left(\frac{r}{z}\right)^2\right]^{5/2}} \cdot \frac{dxdy}{z^2}. \tag{3.25}$$

Substituting $r^2 = x^2+y^2$ into the above equation and integrating it with respect to the whole area of the underside of a foundation, we see that the additional stress induced by the vertical uniform load at depth $z$ under the corner $O$ of the underside of the rectangular foundation will be obtained, as expressed by

$$\begin{aligned}\sigma_z &= \int_o^b \int_o^l \frac{3p}{2\pi} \cdot \frac{z^3 dxdy}{(\sqrt{x^2+y^2+z^2})^5} \\ &= \frac{p}{2\pi}\left[\frac{mn}{\sqrt{1+m^2+n^2}} \cdot \left(\frac{1}{m^2+n^2} + \frac{1}{1+n^2}\right)\right. \\ &\quad \left. + \arctan\left(\frac{m}{n\sqrt{1+m^2+n^2}}\right)\right] \\ &= \alpha_c p, \end{aligned} \tag{3.26}$$

where $\alpha_c$ is the coefficient of the additional stress under the corner $O$ of the underside of a rectangular foundation due to a vertical uniform load. It is a function of $m(= l/b)$ and $n(= z/b)$, namely,

$$\begin{aligned}\alpha_c = f(m,n) = \frac{p}{2\pi}&\left[\frac{mn}{\sqrt{1+m^2+n^2}} \cdot \left(\frac{1}{m^2+n^2} + \frac{1}{1+n^2}\right)\right. \\ &\left. + \arctan\left(\frac{m}{n\sqrt{1+m^2+n^2}}\right)\right]\end{aligned}$$

and it can be read off in Table 3.2, where $l$ is the length of the longer side of the underside of the foundation and $b$ is the width of the shorter side of the underside of the foundation.

For any point inside and outside the range of underside of the foundation, the additional stress can be calculated by using Eq. (3.26) and the principle of superposition.

#### 3.5.2.1 *The point is in the side of the foundation*

Consider a vertical uniform load $p$ acting on the underside of a rectangular foundation *bhfc*, as shown in Fig. 3.16(a). In order to find the additional stress $\sigma_z$ at any depth $z$ under point $M'$, we can draw one auxiliary line $eM'$ parallel to the side of the foundation. Point $M'$ is the common corner of two rectangles $hbM'e$(I) and $eM'fc$(II). Therefore, the additional stress at any depth $z$ under $M'$ is the

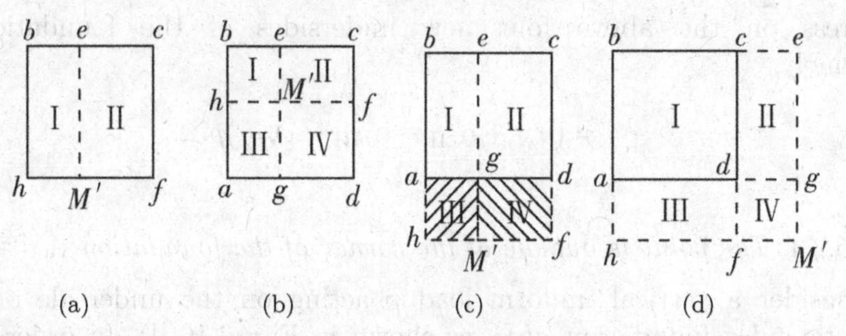


Fig. 3.16. Point that is considered for calculation inside or outside of the underside of the foundation.

sum of the additional stress on the above two new undersides of the foundation, namely,

$$\sigma_z = (\alpha_{c\mathrm{I}} + \alpha_{c\mathrm{II}})p.$$

3.5.2.2 *The point is inside of the foundation*

Consider a vertical uniform load $p$ acting on the underside of a rectangular foundation *abcd*, as shown in Fig. 3.16(b). In order to find the additional stress $\sigma_z$ at any depth $z$ under point $M'$, we can draw two auxiliary lines *eg* and *hf* parallel to the longer side and the shorter side of the foundation, respectively. Point $M'$ is the common corner of four rectangles $bhM'e$(I), $eM'fc$(II), $hagM'$(III), and $M'gdf$(IV). Therefore, the additional stress at any depth $z$ under $M'$ is the sum of the additional stress on the above four new undersides of the foundation, namely,

$$\sigma_z = (\alpha_{c\mathrm{I}} + \alpha_{c\mathrm{II}} + \alpha_{c\mathrm{III}} + \alpha_{c\mathrm{IV}})p.$$

3.5.2.3 *The point is outside of the side of the foundation*

Consider a vertical uniform load $p$ acting on the underside of a rectangular foundation *abcd*, as shown in Fig. 3.16(c). In order to find the additional stress $\sigma_z$ at any depth $z$ under point $M'$, we can draw two auxiliary lines $eM'$ and *hf* parallel to the longer side and the shorter side of the foundation, respectively. Point $M'$ is the common corner of four rectangles $bhM'e$(I), $eM'fc$(II), $hagM'$(III), and $M'gdf$(IV). Therefore, the additional stress at any depth $z$ under $M'$ is the sum of the additional stress on the above four new undersides of the foundation, namely,

$$\sigma_z = (\alpha_{cI} + \alpha_{cII} - \alpha_{cIII} - \alpha_{cIV})p.$$

3.5.2.4 *The point is outside of the corner of the foundation*

Consider a vertical uniform load $p$ acting on the underside of a rectangular foundation *abcd*, as shown in Fig. 3.16(d). In order to find the additional stress $\sigma_z$ at any depth $z$ under point $M'$, we can draw two auxiliary lines $eM'$ and $hM'$ parallel to the longer side and the shorter side of the foundation, respectively. Point $M'$ is the common corner of four rectangles $bhM'e$(I), $eM'fc$(II), $hagM'$(III), and $M'fdg$(IV). Therefore, the additional stress at any depth $z$ under $M'$ is the sum of the additional stress on the above four new undersides of the foundation, namely,

$$\sigma_z = (\alpha_{cI} - \alpha_{cII} - \alpha_{cIII} + \alpha_{cIV})p,$$

where $l$ is the longer side and $b$ is the shorter side in Table 3.3.

**Example 3.3.** There are two adjacent foundations $A$ and $B$, their sizes, positions, and the additional stress distributions are all shown in Fig. 3.17. Considering the effect of the adjacent foundation $B$, try to find the additional stress at a depth $z$ of 2 m under the center point $O$ of foundation $A$.

**Solution.**

- **Step 1:** The additional stress on the center point $O$ due to the vertical uniform load of the foundation $A$.

  In order to obtain the additional stress under point $O$, the underside of foundation $A$ is divided into four rectangles of equal area of 1 m × 1 m. The sum of the additional stress of the four rectangles is the same as the additional stress on the center point $O$ of the foundation $A$, namely,

$$\sigma_z = 4\alpha_c p_A.$$

  With $l/b = 1/1 = 1$ and $z/b = 2/1 = 2$, and from Table 3.3, we can get $\alpha_c = 0.0840$. Therefore, the additional stress under

Table 3.3. Additional stress coefficient $\alpha_c$ under the corner of the underside of rectangular foundation due to a vertical load.

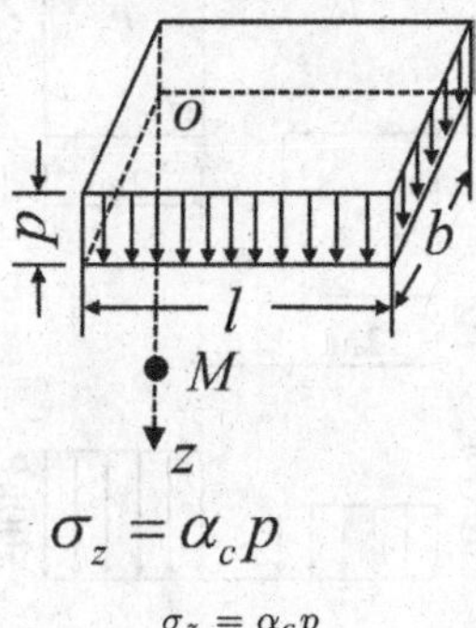


$\sigma_z = \alpha_c p$

| | $l/b$ | | | | | | | | | | |
|---|---|---|---|---|---|---|---|---|---|---|---|
| $z/b$ | 1.0 | 1.2 | 1.4 | 1.6 | 1.8 | 2.0 | 3.0 | 4.0 | 5.0 | 6.0 | 10.0 |
| 0.0 | 0.2500 | 0.2500 | 0.2500 | 0.2500 | 0.2500 | 0.2500 | 0.2500 | 0.250 | 0.2500 | 0.2500 | 0.2500 |
| 0.2 | 0.2486 | 0.2489 | 0.2490 | 0.2491 | 0.2491 | 0.2491 | 0.2492 | 0.2492 | 0.2492 | 0.2492 | 0.2492 |
| 0.4 | 0.2401 | 0.2420 | 0.2429 | 0.2434 | 0.2437 | 0.2439 | 0.2442 | 0.2443 | 0.2443 | 0.2443 | 0.2443 |
| 0.6 | 0.2229 | 0.2275 | 0.2300 | 0.2315 | 0.2324 | 0.2329 | 0.2339 | 0.2341 | 0.2342 | 0.2342 | 0.2342 |
| 0.8 | 0.1999 | 0.2075 | 0.2120 | 0.2147 | 0.2165 | 0.2176 | 0.2196 | 0.2200 | 0.2202 | 0.2202 | 0.2202 |
| 1.0 | 0.1752 | 0.1851 | 0.1911 | 0.1955 | 0.1981 | 0.1999 | 0.2034 | 0.2042 | 0.2044 | 0.2045 | 0.2046 |
| 1.2 | 0.1516 | 0.1626 | 0.1705 | 0.1758 | 0.1793 | 0.1818 | 0.1870 | 0.1882 | 0.1885 | 0.1887 | 0.1888 |
| 1.4 | 0.1308 | 01423 | 0.1508 | 0.1569 | 0.1613 | 0.1644 | 0.1712 | 0.1730 | 0.1735 | 0.1738 | 0.1740 |
| 1.6 | 0.1123 | 0.1241 | 0.1329 | 0.1396 | 0.1445 | 0.1482 | 0.1567 | 0.1590 | 0.1598 | 0.1601 | 0.1604 |
| 1.8 | 0.0969 | 0.1083 | 0.1172 | 0.1241 | 0.1294 | 0.1334 | 0.1434 | 0.1463 | 0.1474 | 0.1478 | 0.1482 |
| 2.0 | 0.0840 | 0.0947 | 0.1034 | 0.1103 | 0.1158 | 0.1202 | 0.1314 | 0.1350 | 0.1363 | 0.1368 | 0.1374 |
| 2.2 | 0.0732 | 0.0832 | 0.0917 | 0.0984 | 0.1039 | 0.1084 | 0.1205 | 0.1248 | 0.1264 | 0.1271 | 0.1277 |
| 2.4 | 0.0642 | 0.0734 | 0.0813 | 0.0879 | 0.0934 | 0.0979 | 0.1108 | 0.1156 | 0.1175 | 0.1184 | 0.1192 |
| 2.6 | 0.0566 | 0.0651 | 0.0725 | 0.0788 | 0.0842 | 0.0887 | 0.1020 | 0.1073 | 0.1095 | 0.1106 | 0.1116 |
| 2.8 | 0.0502 | 0.0580 | 0.0649 | 0.0709 | 0.0761 | 0.0805 | 0.0942 | 0.0999 | 0.1024 | 0.1036 | 0.1048 |
| 3.0 | 0.0447 | 0.0519 | 0.0583 | 0.0640 | 0.0690 | 0.0732 | 0.0870 | 0.0931 | 0.0959 | 0.0973 | 0.0987 |
| 3.2 | 0.0401 | 0.0467 | 0.0526 | 0.0580 | 0.0627 | 0.0668 | 0.0806 | 0.0870 | 0.0900 | 0.096 | 0.0933 |
| 3.4 | 0.0361 | 0.0421 | 0.0477 | 0.0527 | 0.0571 | 0.0611 | 0.0747 | 0.0814 | 0.847 | 0.0864 | 0.0882 |
| 3.6 | 0.0326 | 0.0382 | 0.0433 | 0.0480 | 0.0523 | 0.0561 | 0.0694 | 0.0763 | 0.0799 | 0.0816 | 0.0837 |
| 3.8 | 0.0296 | 0.0348 | 0.0395 | 0.0439 | 0.0479 | 0.0516 | 0.0646 | 0.0717 | 0.0753 | 0.0773 | 0.0796 |
| 4.0 | 0.0270 | 0.0318 | 0.0362 | 0.0404 | 0.0441 | 0.0474 | 0.0603 | 0.0674 | 0.0712 | 0.0733 | 0.0758 |
| 4.2 | 0.0247 | 0.0291 | 0.0333 | 0.0371 | 0.0407 | 0.0439 | 0.0563 | 0.0634 | 0.0674 | 0.0696 | 0.0724 |
| 4.4 | 0.0227 | 0.0268 | 0.0306 | 0.0343 | 0.0376 | 0.0407 | 0.0527 | 0.0597 | 0.0639 | 0.0662 | 0.0692 |
| 4.6 | 0.0209 | 0.0247 | 0.0283 | 0.0317 | 0.0348 | 0.0378 | 0.0493 | 0.0564 | 0.0606 | 0.0630 | 0.0663 |
| 4.8 | 0.0193 | 0.0229 | 0.0262 | 0.0294 | 0.0324 | 0.0352 | 0.0463 | 0.0533 | 0.0576 | 0.0601 | 0.0635 |
| 5.0 | 0.0179 | 0.0212 | 0.0243 | 0.0274 | 0.0302 | 0.0328 | 0.0435 | 0.0504 | 0.0547 | 0.0573 | 0.0610 |
| 6.0 | 0.0127 | 0.0151 | 0.0174 | 0.0196 | 0.02218 | 0.0238 | 0.0325 | 0.0388 | 0.0431 | 0.0460 | 0.0506 |
| 7.0 | 0.0094 | 0.0112 | 0.0130 | 0.0147 | 0.0164 | 0.0180 | 0.0251 | 0.0306 | 0.0346 | 0.0376 | 0.0428 |
| 8.0 | 0.0073 | 0.0087 | 0.0101 | 0.0114 | 0.0127 | 0.0140 | 0.0198 | 0.0246 | 0.0283 | 0.0311 | 0.0367 |
| 9.0 | 0.0058 | 0.0069 | 0.0080 | 0.0091 | 0.0102 | 0.0112 | 0.0161 | 0.0202 | 0.0235 | 0.0262 | 0.0319 |
| 10.0 | 0.0047 | 0.0056 | 0.0065 | 0.0074 | 0.0083 | 0.0092 | 0.0132 | 0.0167 | 0.0198 | 0.0222 | 0.0280 |

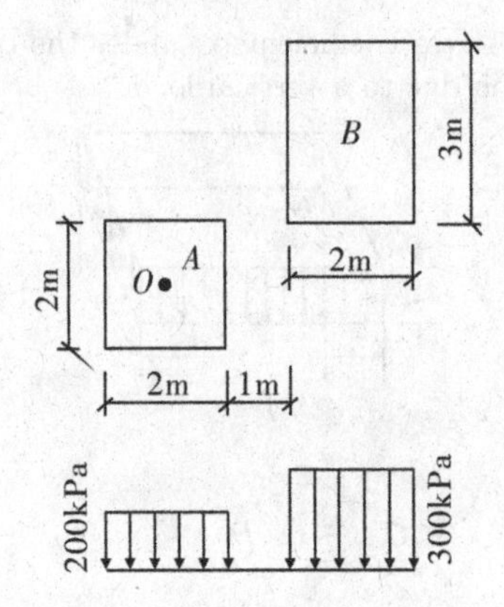


Fig. 3.17. Diagram for Example 3.3.

point $O$ is

$$\sigma_z = 4\alpha_c p_A = 4 \times 0.0840 \times 200 = 67.2 \text{ kPa}.$$

- **Step 2:** The additional stress on the center point $O$ due to the vertical uniform load of the foundation $B$.

  According to the figure, the point $O$ lies outside of the foundation $B$and the additional stress is calculated as shown in Fig. 3.18, that is

  $$\sigma_z = (\alpha_{c\text{I}} - \alpha_{c\text{II}} - \alpha_{c\text{III}} + \alpha_{c\text{IV}})p_B,$$

  where

  rectangle I: $l = 4\,\text{m}$, $b = 4\,\text{m}$; with $l/b = 1$, $z/b = 0.5$, from Table 3.3, we can get $\alpha_c = 0.2315$;
  rectangle II: $l = 4\,\text{m}$, $b = 2\,\text{m}$; with $l/b = 2$, $z/b = 1$, from Table 3.3, we can get $\alpha_c = 0.1999$;
  rectangle III: $l = 4\,\text{m}$, $b = 1\,\text{m}$; with $l/b = 4$, $z/b = 2$, from Table 3.3, we can get $\alpha_c = 0.1350$;
  rectangle IV: $l = 2\,\text{m}$, $b = 1\,\text{m}$; with $l/b = 2$, $z/b = 2$, from Table 3.3, we can get $\alpha_c = 0.1202$.

  $\sigma_z = (\alpha_{c\text{I}} - \alpha_{c\text{II}} - \alpha_{c\text{III}} + \alpha_{c\text{IV}})p_B = (0.2315 - 0.1999 - 0.1350 + 0.1202) \times 300 = 5.28\,\text{kPa}$.

- **Step 3:** The additional stress on the center point $O$ at the 2 m depth of the foundation $A$

  $$\sigma_z = 67.2 + 5.28 = 72.48\,\text{kPa}.$$

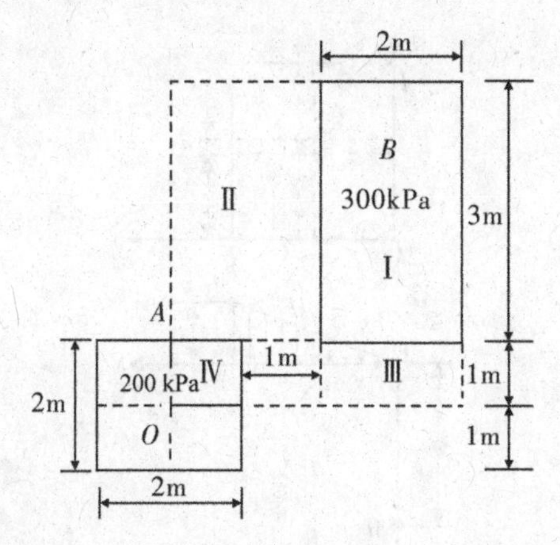


Fig. 3.18. The additional stress on the center point $O$ of the foundation $A$.

### 3.5.3 *Additional stress under the corners of the underside of a rectangular foundation due to a vertical triangular load*

When a triangularly distributed load (i.e. the triangularly distributed compressive pressure, ditto) is applied to the underside of a rectangular foundation, the additional stress under the corner where the load intensity is zero can also be calculated by integrating Eq. (3.24) with respect to the whole loading area. If the maximum intensity of the triangular load on the underside of the rectangular foundation is $p_t$, then the acting force $\mathrm{d}F$ on the infinitesimal area $dxdy$ equals $\frac{p_t x}{b}dxdy$, which can be considered as a concentrated point load, as shown in Fig. 3.19. Therefore, the additional stress at an arbitrary depth $z$ under the corner $O$ induced by this point load can be calculated as given by

$$d\sigma_z = \frac{3p_t}{2\pi b} \cdot \frac{1}{\left[1 + \left(\frac{r}{z}\right)^2\right]^{5/2}} \cdot \frac{xdxdy}{z^2}, \tag{3.27}$$

Substituting $r^2 = x^2 + y^2$ into the above equation and integrating it with respect the whole area of the underside of the foundation, the additional stress induced by the vertical uniform load at depth $z$ under the corner $O$ of the rectangular foundation will be obtained,

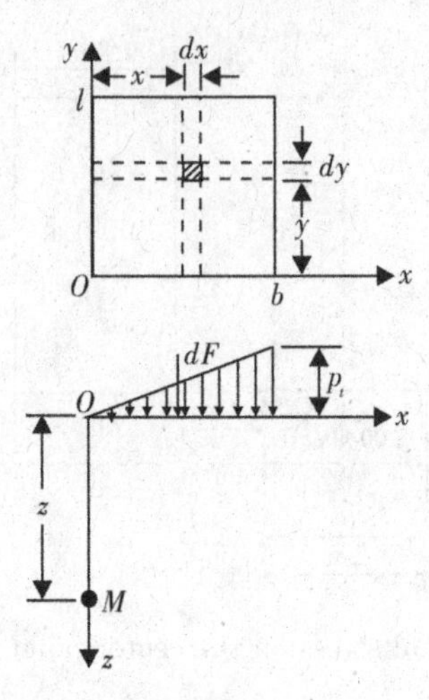


Fig. 3.19. Rectangular foundation subjected to a triangularly distributed load.

as expressed by

$$\sigma_z = \alpha_{tc} p_t, \tag{3.28}$$

$$\alpha_{tc} = \frac{mn}{2\pi}\left[\frac{1}{\sqrt{m^2+n^2}} - \frac{n^2}{(1+n^2)\sqrt{1+m^2+n^2}}\right], \tag{3.29}$$

where $\alpha_{tc}$ is the coefficient of the additional stress under the corner of the underside of a rectangular foundation due to a vertical triangular load. It is a function of $m(= l/b)$ and $n(= z/b)$ and it can be read off in Table 3.4, where $b$ is the length of one side of the underside of the foundation in the loading variation direction and $l$ is the length of another side.

For the additional stress at any point inside and outside the range of the underside of the foundation, it can also be calculated using the principle of superposition. Two points should, however, be noted: the calculation point should be on the vertical line under the point where the triangular load intensity is zero; $b$ represents the side length of the underside of the rectangular foundation in the loading variation direction.

When a horizontal uniform load $p_h$ is applied to the underside of a rectangular foundation (Fig. 3.20), the additional stress at an arbitrary depth $z$ under the corner of the foundation is given as

$$\sigma_z = \pm\alpha_l \cdot p_h, \tag{3.30a}$$

$$\alpha_l = \frac{m}{2\pi}\left[\frac{1}{\sqrt{m^2+n^2}} - \frac{n^2}{(1+n^2)\sqrt{1+m^2+n^2}}\right], \tag{3.30b}$$

Table 3.4. Additional stress coefficient $\alpha_{tc}$ under the corner of the underside of a rectangular foundation due to a vertical triangular load.

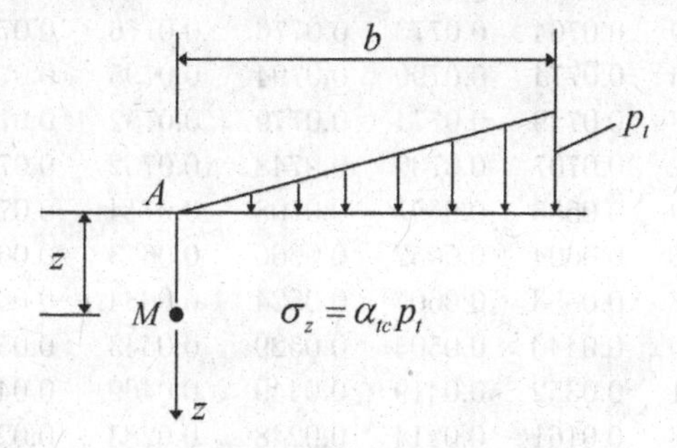


| | $l/b$ | | | | | | | |
|---|---|---|---|---|---|---|---|---|
| $z/b$ | 0.2 | 0.4 | 0.6 | 0.8 | 1.0 | 1.2 | 1.4 | 1.6 |
| 0.0 | 0.0000 | 0.0000 | 0.0000 | 0.0000 | 0.0000 | 0.0000 | 0.0000 | 0.0000 |
| 0.2 | 0.0223 | 0.0280 | 0.0296 | 0.0301 | 0.0304 | 0.0305 | 0.0305 | 0.0306 |
| 0.4 | 0.0269 | 0.0420 | 0.0487 | 0.0517 | 0.0531 | 0.0539 | 0.0543 | 0.0545 |
| 0.6 | 0.0259 | 0.0448 | 0.0560 | 0.0621 | 0.0654 | 0.0673 | 0.0684 | 0.0690 |
| 0.8 | 0.0232 | 0.0421 | 0.0553 | 0.0637 | 0.0688 | 0.0720 | 0.0739 | 0.0751 |
| 1.0 | 0.0201 | 0.0375 | 0.0508 | 0.0602 | 0.0666 | 0.0708 | 0.0735 | 0.0753 |
| 1.2 | 0.0171 | 0.0324 | 0.0450 | 0.0546 | 0.0615 | 0.0664 | 0.0698 | 0.0721 |
| 1.4 | 0.0145 | 0.0278 | 0.0392 | 0.0483 | 0.0554 | 0.0606 | 0.0644 | 0.0672 |
| 1.6 | 0.0123 | 0.0238 | 0.0339 | 0.0424 | 0.0492 | 0.0545 | 0.0586 | 0.0616 |
| 1.8 | 0.0105 | 0.0204 | 0.0294 | 0.0371 | 0.0435 | 0.0487 | 0.0528 | 0.0560 |
| 2.0 | 0.0090 | 0.0176 | 0.0255 | 0.0324 | 0.0384 | 0.0434 | 0.0474 | 0.0507 |
| 2.5 | 0.0063 | 0.0125 | 0.0183 | 0.0236 | 0.0284 | 0.0326 | 0.0362 | 0.0393 |
| 3.0 | 0.0046 | .0.0092 | 0.0135 | 0.0176 | 0.0214 | 0.0249 | 0.0280 | 0.0307 |
| 5.0 | 0.0018 | 0.0036 | 0.0054 | 0.0071 | 0.0088 | 0.0104 | 0.0120 | 0.0135 |
| 7.0 | 0.0009 | 0.0019 | 0.0028 | 0.0038 | 0.0047 | 0.0056 | 0.0064 | 0.0073 |
| 10.0 | 0.0005 | 0.0009 | 0.0014 | 0.0019 | 0.0023 | 0.0028 | 0.0033 | 0.0037 |

(*Continued*)

Table 3.4. (*Continued*)

| | $l/b$ | | | | | | |
|---|---|---|---|---|---|---|---|
| $z/b$ | 1.8 | 2.0 | 3.0 | 4.0 | 6.0 | 8.0 | 10.0 |
| 0.0 | 0.0000 | 0.0000 | 0.0000 | 0.0000 | 0.0000 | 0.0000 | 0.0000 |
| 0.2 | 0.0306 | 0.0306 | 0.0306 | 0.0306 | 0.0306 | 0.0306 | 0.0306 |
| 0.4 | 0.0546 | 0.0547 | 0.0548 | 0.0549 | 0.0549 | 0.0549 | 0.0549 |
| 0.6 | 0.0694 | 0.0696 | 0.0701 | 0.0702 | 0.0702 | 0.0702 | 0.0702 |
| 0.8 | 0.0759 | 0.0764 | 0.0773 | 0.0776 | 0.0776 | 0.0776 | 0.0776 |
| 1.0 | 0.0766 | 0.0774 | 0.0790 | 0.0794 | 0.0795 | 0.0796 | 0.0796 |
| 1.2 | 0.0738 | 0.0749 | 0.0774 | 0.0779 | 0.0782 | 0.0783 | 0.0783 |
| 1.4 | 0.0692 | 0.0707 | 0.0739 | 0.0748 | 0.0752 | 0.0752 | 0.0753 |
| 1.6 | 0.0639 | 0.0656 | 0.0697 | 0.0708 | 0.0714 | 0.0715 | 0.0715 |
| 1.8 | 0.0585 | 0.0604 | 0.0652 | 0.0666 | 0.0673 | 0.0675 | 0.0675 |
| 2.0 | 0.0533 | 0.0553 | 0.0607 | 0.0624 | 0.0634 | 0.0636 | 0.0636 |
| 2.5 | 0.0419 | 0.0440 | 0.0504 | 0.0529 | 0.0543 | 0.0547 | 0.0548 |
| 3.0 | 0.0331 | 0.0352 | 0.0419 | 0.0449 | 0.0469 | 0.0474 | 0.0476 |
| 5.0 | 0.0148 | 0.0161 | 0.0214 | 0.0248 | 0.0283 | 0.0296 | 0.0301 |
| 7.0 | 0.0081 | 0.0089 | 0.0124 | 0.0152 | 0.01860 | 0.0204 | 0.0212 |
| 10.0 | 0.0041 | 0.0046 | 0.0066 | 0.0084 | 0.0111 | 0.0128 | 0.0139 |

where $\alpha_l$ is the coefficient of the additional stress under the corner of the underside of a rectangular foundation due to a horizontal uniform load. It is a function of $m(= l/b)$ and $n(= z/b)$ and it can be read off in Table 3.5, where $b$ is the length of one side of the underside of the foundation in the direction parallel to the horizontal load direction and $l$ is the length of another side.

In the above equation, the positive sign "+" is taken at the calculation point under the stopping end of the horizontal uniform load (under point 2) and the negative sign "–" is taken at the calculation point under the starting end of the horizontal uniform load (under point 1).

The additional stress at any point inside and outside the range of the underside of the foundation can also be calculated using the principle of superposition.

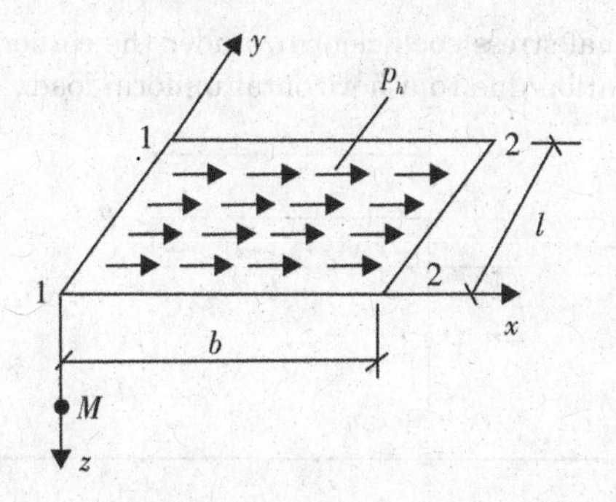


Fig. 3.20. Rectangular foundation subjected to a horizontal uniform load.

## 3.6 Additional Stress in Plane Problem

In theory, when the ratio of length $l$ to width $b$ of a foundation approaches an infinitely large value, the stress state in the interior of the ground is a plane problem. In engineering practice, however, there is no infinitely long foundation in reality. Based on various researches, the stress state of foundations with a ratio $l/b$ greater than or equal to 10 is similar to those foundations with a ratio $l/b$ approaching an infinitely large value. The discrepancy between the two cases is acceptable from an engineering point of view. Sometimes, the stress state for foundations with $l/b$ greater than 5 is also calculated in accordance with the plane problem.

### 3.6.1 *Stress increment due to a vertical line load*

Vertical line load is defined as the vertical uniform load applied on an infinitely long line, as shown in Fig. 3.21. When a vertical line load is applied on the ground surface, the stress increment at an arbitrary depth $z$ at the interior of the ground can be calculated using the Flamant solution, as given by

$$\sigma_z = \frac{2pz^3}{\pi(x^2+z^2)^2}, \tag{3.31a}$$

$$\sigma_x = \frac{2px^2z}{\pi(x^2+z^2)^2}, \tag{3.31b}$$

Table 3.5. Additional stress coefficient $\alpha_l$ under the corner of the underside of a rectangular foundation due to a horizontal uniform load.

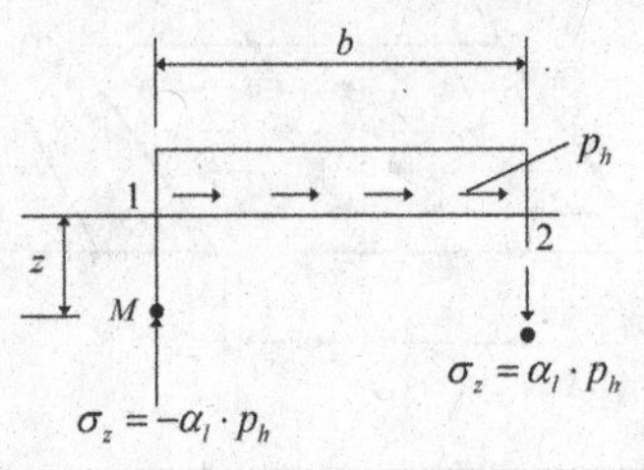


| | $l/b$ | | | | | | | | | |
|---|---|---|---|---|---|---|---|---|---|---|
| $z/b$ | 0.2 | 0.4 | 1.0 | 1.6 | 2.0 | 3.0 | 4.0 | 6.0 | 8.0 | 10.0 |
| 0.0 | 0.1592 | 0.1592 | 0.1592 | 0.1592 | 0.1592 | 0.1592 | 0.1592 | 0.1592 | 0.1592 | 0.1592 |
| 0.2 | 0.1114 | 0.1401 | 0.1518 | 0.1528 | 0.1529 | 0.1530 | 0.1530 | 0.1530 | 0.1530 | 0.1530 |
| 0.4 | 0.0672 | 0.1049 | 0.1328 | 0.1362 | 0.1367 | 0.1371 | 0.1372 | 0.1372 | 0.1372 | 0.1372 |
| 0.6 | 0.0432 | 0.0746 | 0.1091 | 0.1150 | 0.1160 | 0.1168 | 0.1169 | 0.1170 | 0.1170 | 0.1170 |
| 0.8 | 0.0290 | 0.0527 | 0.0861 | 0.0939 | 0.0955 | 0.0967 | 0.0969 | 0.0970 | 0.0970 | 0.0970 |
| 1.0 | 0.0201 | 0.0375 | 0.0666 | 0.0753 | 0.0774 | 0.0790 | 0.0794 | 0.0795 | 0.0796 | 0.0796 |
| 1.2 | 0.0142 | 0.0270 | 0.0512 | 0.0601 | 0.0624 | 0.0645 | 0.0650 | 0.0652 | 0.0652 | 0.0652 |
| 1.4 | 0.0103 | 0.0199 | 0.0395 | 0.0480 | 0.0505 | 0.0528 | 0.0534 | 0.0537 | 0.0537 | 0.0538 |
| 1.6 | 0.0077 | 0.0149 | 0.0308 | 0.0385 | 0.0410 | 0.0436 | 0.0443 | 0.0446 | 0.0447 | 0.0447 |
| 1.8 | 0.0058 | 0.0113 | 0.0242 | 0.0311 | 0.0336 | 0.0362 | 0.0370 | 0.0374 | 0.0375 | 0.0375 |
| 2.0 | 0.0045 | 0.0088 | 0.0192 | 0.0253 | 0.0277 | 0.0303 | 0.0312 | 0.0317 | 0.0318 | 0.0318 |
| 2.5 | 0.0025 | 0.0050 | 0.0113 | 0.0157 | 0.0176 | 0.0202 | 0.0211 | 0.0217 | 0.0219 | 0.0219 |
| 3.0 | 0.0015 | 0.0031 | 0.0071 | 0.0102 | 0.0117 | 0.0140 | 0.0150 | 0.0156 | 0.0158 | 0.0159 |
| 5.0 | 0.0004 | 0.0007 | 0.0018 | 0.0027 | 0.0032 | 0.0043 | 0.0050 | 0.0057 | 0.0059 | 0.0062 |
| 7.0 | 0.0001 | 0.0003 | 0.0007 | 0.0010 | 0.0013 | 0.0018 | 0.0022 | 0.0027 | 0.0029 | 0.0030 |
| 10.0 | 0.0001 | 0.0001 | 0.0002 | 0.0004 | 0.0005 | 0.0007 | 0.0008 | 0.0011 | 0.0013 | 0.0014 |

$$\tau_{zx} = \tau_{xz} = \frac{2pxz^2}{\pi(x^2 + z^2)^2}, \tag{3.31c}$$

where $p$ denotes the line load on a unit length (kN/m); $x$ and $z$ are coordinates of the calculation point.

### 3.6.2 *Additional stress beneath the underside of the a strip foundation due to a vertical uniform load*

When a vertical uniform load of intensity $p$ is applied to the underside of a foundation as shown in Fig. 3.22, the additional stress at

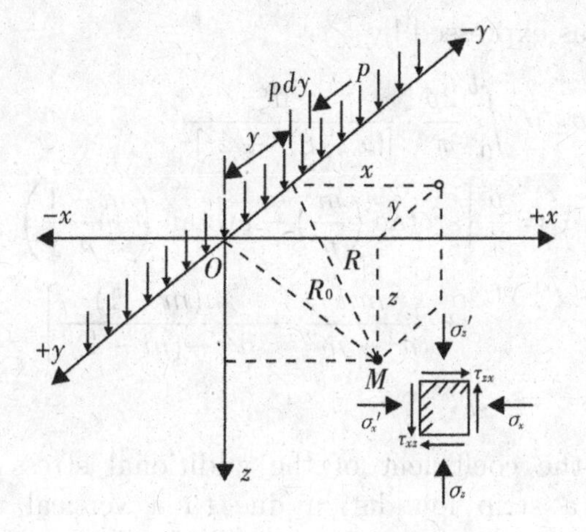


Fig. 3.21. Stress state of the soil due to a vertical line load.

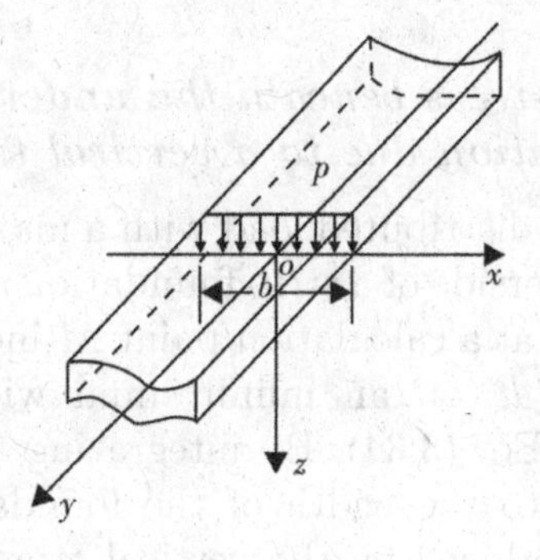


Fig. 3.22. Strip foundation subjected to a vertical uniform load.

an arbitrary point $M$ induced by the line load $d\bar{p} = pd\xi$ on an infinitesimal width $d\xi$ can first be calculated by using Eq. (3.32), as given by

$$d\sigma_z = \frac{2p}{\pi} \cdot \frac{z^3 d\xi}{[(x-\xi)^2 + z^2]^2}. \tag{3.32}$$

The additional stress beneath the underside of the strip foundation due to a vertical uniform load can then be calculated by integrating the above equation with respect to the width of the loading area, as expressed by

$$\sigma_z = \int_0^b \frac{2p}{\pi} \cdot \frac{z^3 d\xi}{[(x-\xi)^2 + z^2]^2}$$
$$= \frac{p}{\pi}\left[\arctan\left(\frac{m}{n}\right) - \arctan\left(\frac{m-1}{n}\right) + \frac{mn}{n^2+m^2} - \frac{n(m-1)}{n^2+(m-1)^2}\right]$$
$$= \alpha_z^s p, \tag{3.33}$$

where $\alpha_z^s$ is the coefficient of the additional stress beneath the underside of a strip foundation due to a vertical uniform load (compressive pressure, ditto). It is a function of $m(= x/b)$ and $n(z/b)$ and it can be read off in Table 3.6, where $b$ is the width of the strip foundation as shown in Fig. 3.22; $x$ and $z$ are coordinates of the calculation point.

### 3.6.3 *Additional stress beneath the underside of a strip foundation due to a vertical triangular load*

When a triangularly distributed load with a maximum intensity of $p_t$ is applied to the underside of a strip foundation as shown in Fig. 3.23, the additional stress at a calculation point $M$ induced by the vertical line load $dp = \frac{p_t}{b}\xi d\xi$ on an infinitesimal width $d\xi$ can first be calculated by using Eq. (3.34). By integrating this additional stress on $d\xi$ with respect to the width of the foundation, the additional stress at point $M$ induced by the vertical triangular load can then be calculated, as given by

$$\sigma_z = \int_0^b \frac{2p_t}{\pi b} \cdot \frac{z^3 \xi d\xi}{[(x-\xi)^2 + z^2]^2}$$
$$= \frac{p_t}{\pi}\left\{m\left[\arctan\left(\frac{m}{n}\right) - \arctan\left(\frac{m-1}{n}\right)\right] - \frac{n(m-1)}{n^2+(m-1)^2}\right\},$$
$$= \alpha_t^s p_t, \tag{3.34}$$

where $\alpha_t^s$ is the coefficient of the additional stress beneath the underside of a strip foundation due to a vertical triangular distributed load (compressive pressure, ditto). It is a function of $m(= x/b)$ and

Table 3.6. Additional stress coefficient $\alpha_z^s$ beneath the underside of a strip foundation due to a vertical uniform load.

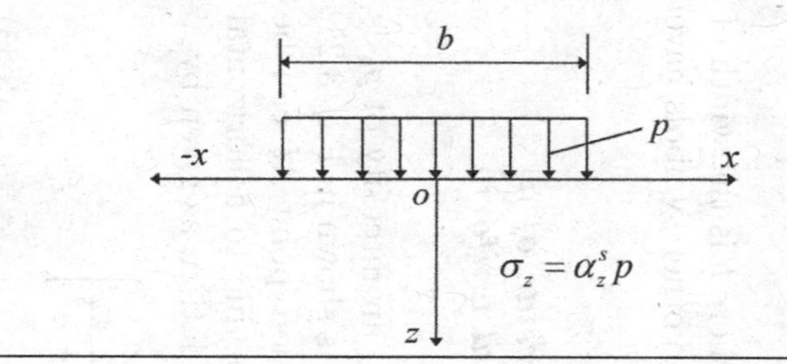


| $z/b$ | $x/b$: 0.0 | 0.10 | 0.25 | 0.35 | 0.50 | 0.75 | 1.00 | 1.50 | 2.00 | 2.50 | 3.00 | 4.00 | 5.00 |
|---|---|---|---|---|---|---|---|---|---|---|---|---|---|
| 0.00 | 1.000 | 1.000 | 1.000 | 1.000 | 0.500 | 0.000 | 0.000 | 0.000 | 0.000 | 0.000 | 0.000 | 0.000 | 0.000 |
| 0.05 | 1.000 | 1.000 | 0.995 | 0.970 | 0.500 | 0.002 | 0.000 | 0.000 | 0.000 | 0.000 | 0.000 | 0.000 | 0.000 |
| 0.10 | 0.997 | 0.996 | 0.986 | 0.965 | 0.499 | 0.010 | 0.005 | 0.000 | 0.000 | 0.000 | 0.000 | 0.000 | 0.000 |
| 0.15 | 0.993 | 0.987 | 0.968 | 0.910 | 0.498 | 0.033 | 0.008 | 0.001 | 0.000 | 0.000 | 0.000 | 0.000 | 0.000 |
| 0.25 | 0.960 | 0.954 | 0.905 | 0.805 | 0.496 | 0.088 | 0.019 | 0.002 | 0.001 | 0.000 | 0.000 | 0.000 | 0.000 |
| 0.35 | 0.907 | 0.900 | 0.832 | 0.732 | 0.492 | 0.148 | 0.039 | 0.006 | 0.003 | 0.001 | 0.000 | 0.000 | 0.000 |
| 0.50 | 0.820 | 0.812 | 0.735 | 0.651 | 0.481 | 0.218 | 0.082 | 0.017 | 0.005 | 0.002 | 0.001 | 0.000 | 0.000 |
| 0.75 | 0.668 | 0.658 | 0.610 | 0.552 | 0.450 | 0.263 | 0.146 | 0.040 | 0.017 | 0.005 | 0.005 | 0.001 | 0.000 |
| 1.00 | 0.552 | 0.541 | 0.513 | 0.475 | 0.410 | 0.288 | 0.185 | 0.071 | 0.029 | 0.013 | 0.007 | 0.002 | 0.001 |
| 1.50 | 0.396 | 0.395 | 0.379 | 0.353 | 0.332 | 0.273 | 0.211 | 0.114 | 0.055 | 0.030 | 0.018 | 0.006 | 0.003 |
| 2.00 | 0.306 | 0.304 | 0.292 | 0.288 | 0.275 | 0.242 | 0.205 | 0.134 | 0.083 | 0.051 | 0.028 | 0.013 | 0.006 |
| 2.50 | 0.245 | 0.244 | 0.239 | 0.237 | 0.231 | 0.215 | 0.188 | 0.139 | 0.098 | 0.065 | 0.034 | 0.021 | 0.010 |
| 3.00 | 0.208 | 0.208 | 0.206 | 0.202 | 0.198 | 0.185 | 0.171 | 0.136 | 0.103 | 0.075 | 0.053 | 0.028 | 0.015 |
| 4.00 | 0.160 | 0.160 | 0.158 | 0.156 | 0.153 | 0.147 | 0.140 | 0.122 | 0.102 | 0.081 | 0.066 | 0.040 | 0.025 |
| 5.00 | 0.126 | 0.126 | 0.125 | 0.125 | 0.124 | 0.121 | 0.117 | 0.107 | 0.095 | 0.082 | 0.069 | 0.046 | 0.034 |

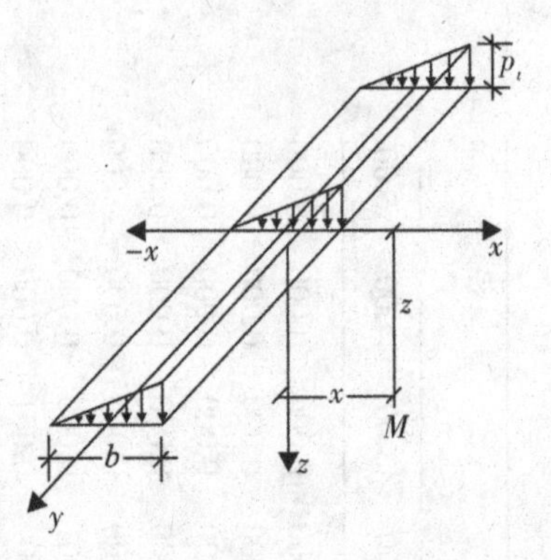


Fig. 3.23. Strip foundation subjected to a vertical triangular distributed load.

$n(= z/b)$ and it can be read off in Table 3.7, where $b$ is the width of the strip foundation as shown in Fig. 3.23. The other symbols have the same meanings as above.

### 3.6.4 *Additional stress beneath the underside of a strip foundation due to a horizontal uniform load*

Similarly, when a horizontal uniform load with an intensity of $p_h$ is applied to the underside of a strip foundation as shown in Fig. 3.24, the additional stress at an arbitrary calculation point $M$ can be calculated by integrating the additional stress due to a horizontal line load with respect to the width of the foundation, as given by

$$\sigma_z = \frac{p_h}{\pi}\left[\frac{n^2}{(m-1)^2+n^2} - \frac{n^2}{m^2+n^2}\right]$$
$$= \alpha_h p, \tag{3.35}$$

where $\alpha_h$ is the coefficient of the additional stress beneath the underside of a strip foundation due to a horizontal uniform load (compressive pressure, ditto). It is a function of $m(= x/b)$ and $n(= z/b)$ and it can be read off in Table 3.8. The other symbols have the same meanings as above.

It must be noted that when calculating the additional stress under the strip foundation, the coordinate system must satisfy the requirements specified separately in Figs. 3.21–3.23.

Table 3.7. Additional stress coefficient $\alpha_t^s$ beneath the underside of a strip foundation due to a vertical triangular distributed load.

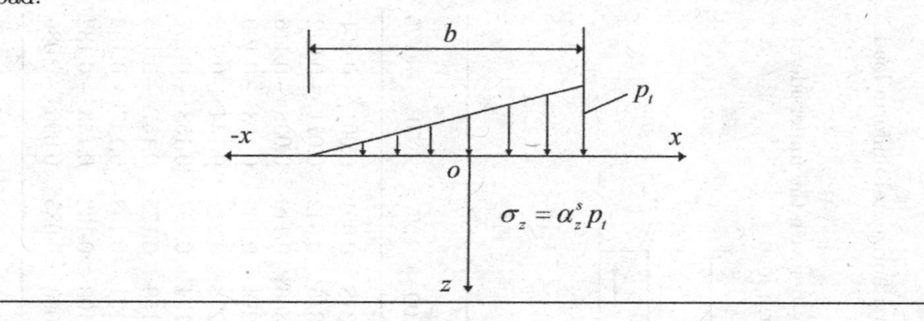


| | x/b | | | | | | | | | | | | | |
|---|---|---|---|---|---|---|---|---|---|---|---|---|---|---|
| z/b | −2.00 | −1.50 | −1.00 | −0.75 | −0.50 | −0.25 | 0.00 | 0.25 | 0.50 | 0.75 | 1.00 | 1.50 | 2.00 | 3.00 |
| 0.00 | 0.00 | 0.00 | 0.00 | 0.00 | 0.00 | 0.25 | 0.50 | 0.75 | 0.50 | 0.00 | 0.00 | 0.00 | 0.00 | 0.00 |
| 0.25 | 0.00 | 0.00 | 0.00 | 0.01 | 0.08 | 0.26 | 0.48 | 0.65 | 0.42 | 0.08 | 0.02 | 0.00 | 0.00 | 0.00 |
| 0.50 | 0.00 | 0.01 | 0.02 | 0.05 | 0.13 | 0.26 | 0.41 | 0.47 | 0.35 | 0.16 | 0.06 | 0.01 | 0.00 | 0.00 |
| 0.75 | 0.01 | 0.01 | 0.05 | 0.08 | 0.15 | 0.25 | 0.33 | 0.36 | 0.29 | 0.19 | 0.10 | 0.03 | 0.01 | 0.00 |
| 1.00 | 0.01 | 0.03 | 0.06 | 0.10 | 0.16 | 0.22 | 0.28 | 0.29 | 0.25 | 0.18 | 0.12 | 0.05 | 0.02 | 0.00 |
| 1.50 | 0.02 | 0.05 | 0.09 | 0.11 | 0.15 | 0.18 | 0.20 | 0.20 | 0.19 | 0.16 | 0.13 | 0.07 | 0.04 | 0.01 |
| 2.00 | 0.03 | 0.06 | 0.09 | 0.11 | 0.14 | 0.16 | 0.16 | 0.16 | 0.15 | 0.13 | 0.12 | 0.08 | 0.05 | 0.02 |
| 2.50 | 0.04 | 0.06 | 0.08 | 0.12 | 0.13 | 0.13 | 0.13 | 0.13 | 0.12 | 0.11 | 0.10 | 0.07 | 0.05 | 0.02 |
| 3.00 | 0.05 | 0.06 | 0.08 | 0.09 | 0.10 | 0.10 | 0.11 | 0.11 | 0.10 | 0.10 | 0.09 | 0.07 | 0.05 | 0.03 |
| 4.00 | 0.05 | 0.06 | 0.07 | 0.07 | 0.08 | 0.08 | 0.08 | 0.08 | 0.08 | 0.08 | 0.07 | 0.06 | 0.05 | 0.03 |
| 5.00 | 0.05 | 0.05 | 0.06 | 0.06 | 0.06 | 0.06 | 0.06 | 0.06 | 0.06 | 0.06 | 0.06 | 0.05 | 0.04 | 0.03 |

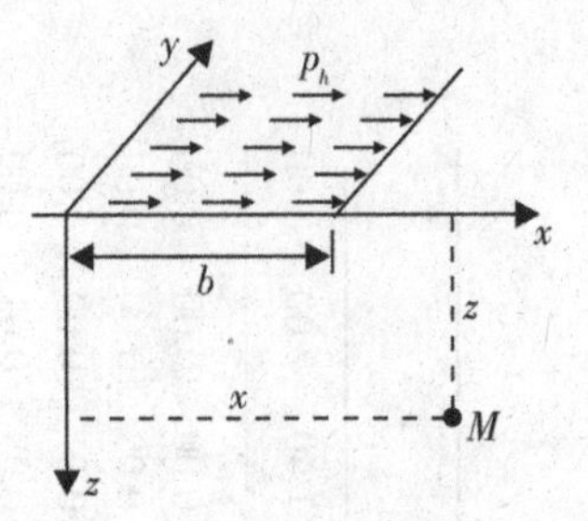


Fig. 3.24. Strip foundation subjected to a horizontal uniform load.

Table 3.8. Additional stress coefficient $\alpha_h$ beneath the underside of a strip foundation due to a horizontal uniform load.

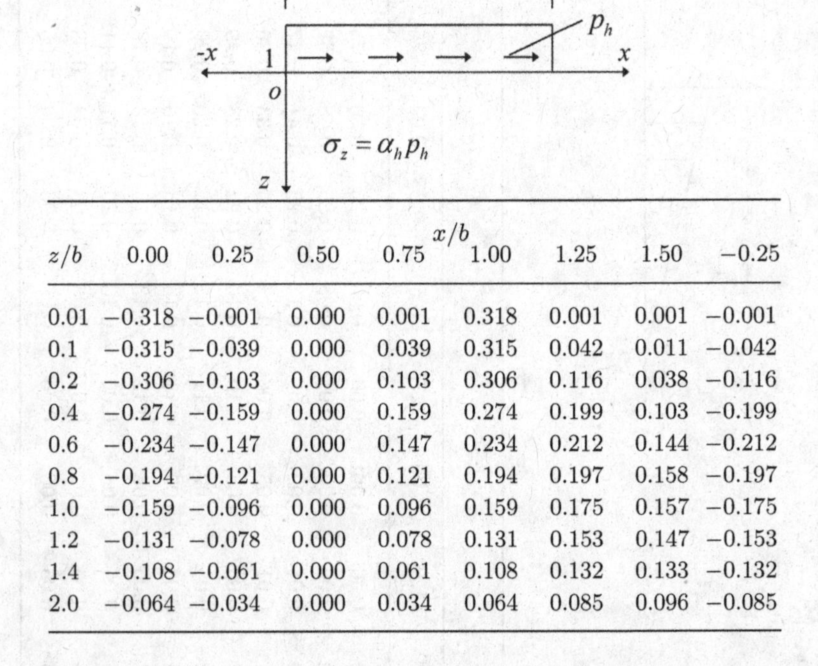


| $z/b$ | 0.00 | 0.25 | 0.50 | 0.75 | 1.00 | 1.25 | 1.50 | −0.25 |
|---|---|---|---|---|---|---|---|---|
| | | | | $x/b$ | | | | |
| 0.01 | −0.318 | −0.001 | 0.000 | 0.001 | 0.318 | 0.001 | 0.001 | −0.001 |
| 0.1 | −0.315 | −0.039 | 0.000 | 0.039 | 0.315 | 0.042 | 0.011 | −0.042 |
| 0.2 | −0.306 | −0.103 | 0.000 | 0.103 | 0.306 | 0.116 | 0.038 | −0.116 |
| 0.4 | −0.274 | −0.159 | 0.000 | 0.159 | 0.274 | 0.199 | 0.103 | −0.199 |
| 0.6 | −0.234 | −0.147 | 0.000 | 0.147 | 0.234 | 0.212 | 0.144 | −0.212 |
| 0.8 | −0.194 | −0.121 | 0.000 | 0.121 | 0.194 | 0.197 | 0.158 | −0.197 |
| 1.0 | −0.159 | −0.096 | 0.000 | 0.096 | 0.159 | 0.175 | 0.157 | −0.175 |
| 1.2 | −0.131 | −0.078 | 0.000 | 0.078 | 0.131 | 0.153 | 0.147 | −0.153 |
| 1.4 | −0.108 | −0.061 | 0.000 | 0.061 | 0.108 | 0.132 | 0.133 | −0.132 |
| 2.0 | −0.064 | −0.034 | 0.000 | 0.034 | 0.064 | 0.085 | 0.096 | −0.085 |

**Example 3.4.** A concentrated load $F = 400$ kN/m and a moment $M = 20$ kN/m are acting on the strip foundation, as shown in Fig. 3.25. Try to find the additional stress on the center of the strip foundation and draw its profile with respect to the depth of soil.

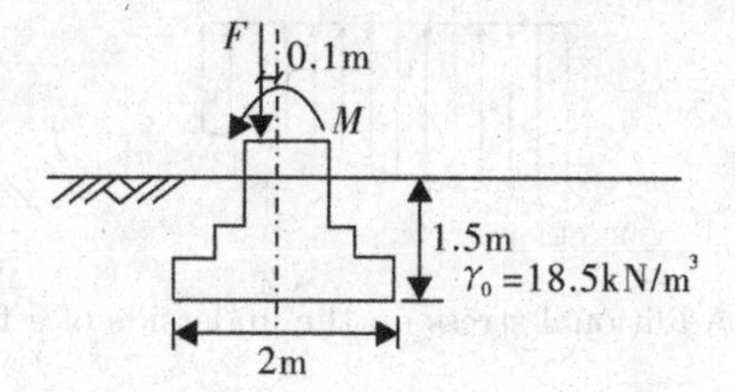


Fig. 3.25. Diagram for Example 3.4.

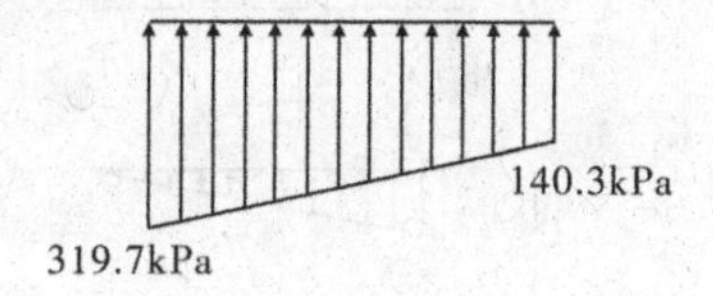


Fig. 3.26. Contact pressure diagram.

**Solution.**

**Step 1:** Calculate the offsetting $e$

$$e = \frac{M + F \times 0.1}{F + G} = \frac{20 + 400 \times 0.1}{400 + 20 \times 2 \times 1 \times 1.5} = 0.13\,\text{m}$$

**Step 2:** Calculate the contact pressure

$$p_{\substack{\max\\ \min}} = \frac{F+G}{b}\left(1 \pm \frac{6e}{b}\right) = \frac{400+60}{2}\left(1 \pm \frac{6 \times 0.13}{2}\right)$$

$$= \begin{matrix} 319.7\,\text{kPa} \\ 140.3\,\text{kPa} \end{matrix}.$$

The contact pressure diagram is shown in Fig. 3.26.

**Step 3:** Calculate the additional stress on the underside of the foundation

$$p_{\substack{0\max\\ 0\min}} = p_{\substack{\max\\ \min}} - \gamma_0 d = \begin{matrix} 319.7 \\ 140.3 \end{matrix} - 18.5 \times 1.5 = \begin{matrix} 292.0\,\text{kPa} \\ 112.6\,\text{kPa} \end{matrix}.$$

Fig. 3.27 shows the additional stress on the underside of a foundation.

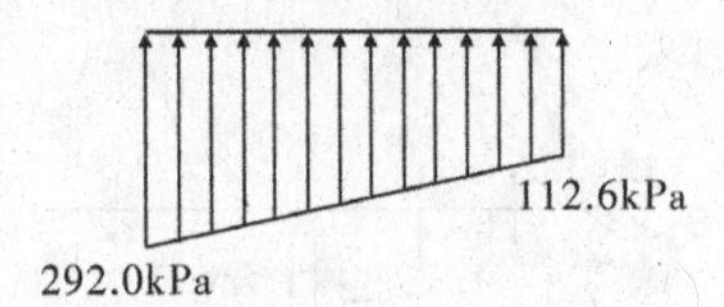


Fig. 3.27. Additional stress on the underside of a foundation.

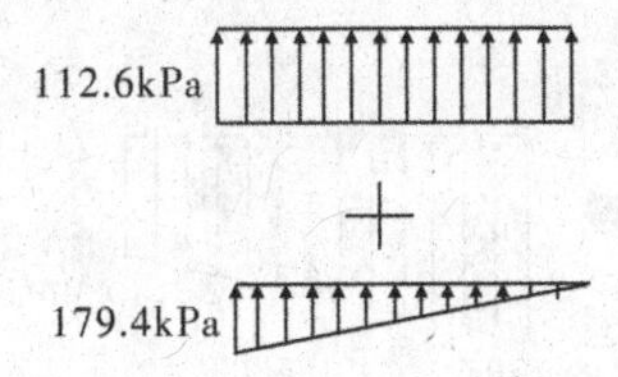


Fig. 3.28. Breakdown of the additional stress on the underside of a foundation.

**Step 4:** Calculate the additional stress at the center of the foundation.

The trapezoidal vertical additional stress on underside of the foundation is decomposed into a vertical uniform additional stress $p_1 = 112.6\,\text{kPa}$ and a vertical triangular additional stress $p_{2t} = 179.4\,\text{kPa}$, as shown in Fig. 3.28. The additional stress on the center of the foundation due to each load can be calculated and the calculation process is shown in Table 3.9. The additional stress coefficients shown in this table are read off in Tables 3.6 and 3.7.

The additional stress on the center of the foundation is calculated by the following equation, namely,

$$\sigma_z = \alpha_z^s p_1 + \alpha_t^s p_{2t}.$$

The additional stress on the center of the foundation diagram is shown in Fig. 3.29.

## Exercises

3.1 A ground soil profile is shown in Fig. 3.30. Find the vertical geostatic stress and plot its distribution curve.

3.2 A rectangular foundation is shown in Fig. 3.31. A vertical load of 800 kN is applied on point $A$ of the foundation. If the eccentricity

Table 3.9. Coefficients of additional stress.

| Types of load | $x$ (m) | $z$ (m) | $b$ (m) | $x/b$ | $z/b$ | Additional stress coefficient | Additional stress (kPa) | Sum of additional stress due to the two loads |
|---|---|---|---|---|---|---|---|---|
| Vertical | 0 | 0 | 2 | 0 | 0 | 1.00 ($\alpha_z^s$) | 112.6 | 202.3 |
| uniform | 0 | 0.5 | 2 | 0 | 0.25 | 0.960 | 108.096 | 194.2 |
| load | 0 | 1 | 2 | 0 | 0.5 | 0.820 | 92.332 | 165.9 |
| | 0 | 2 | 2 | 0 | 1 | 0.552 | 62.155 | 112.4 |
| | 0 | 3 | 2 | 0 | 1.5 | 0.396 | 44.59 | 80.5 |
| | 0 | 4 | 2 | 0 | 2 | 0.306 | 34.456 | 63.2 |
| Vertical | 0 | 0 | 2 | 0 | 0 | 0.50 ($\alpha_t^s$) | 89.7 | |
| triangular | 0 | 0.5 | 2 | 0 | 0.25 | 0.48 | 86.112 | |
| distributed | 0 | 1 | 2 | 0 | 0.5 | 0.41 | 73.554 | |
| load | 0 | 2 | 2 | 0 | 1 | 0.28 | 50.232 | |
| | 0 | 3 | 2 | 0 | 1.5 | 0.20 | 35.88 | |
| | 0 | 4 | 2 | 0 | 2 | 0.16 | 28.704 | |

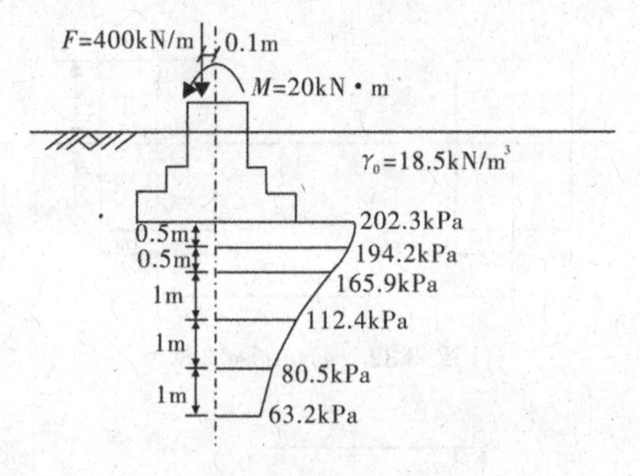


Fig. 3.29. Additional stress on the center of the foundation diagram.

is 0.5 m, try to find the additional stress at a depth 5 m under point $B$ in the figure.

3.3 The two foundation sizes are shown in Fig. 3.32. A vertical centric load 1940 kN is applied on the foundations I and II. The embedded depths of both the foundations are 1.5 m. Find the

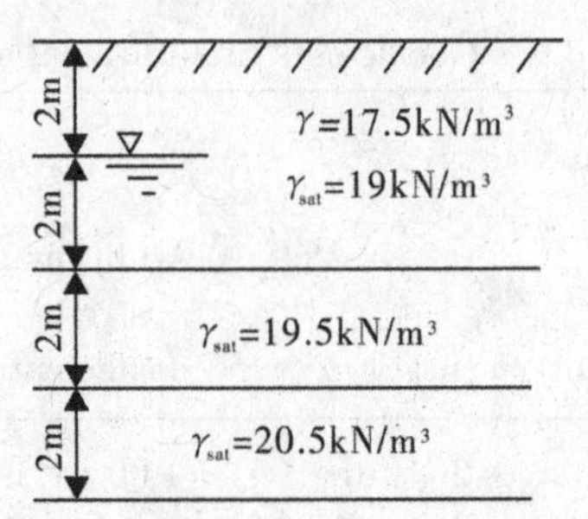


Fig. 3.30. Exercise 3.1.

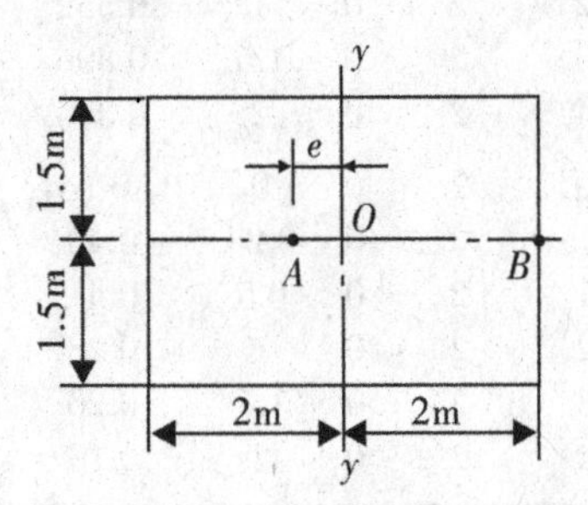


Fig. 3.31. Exercise 3.2.

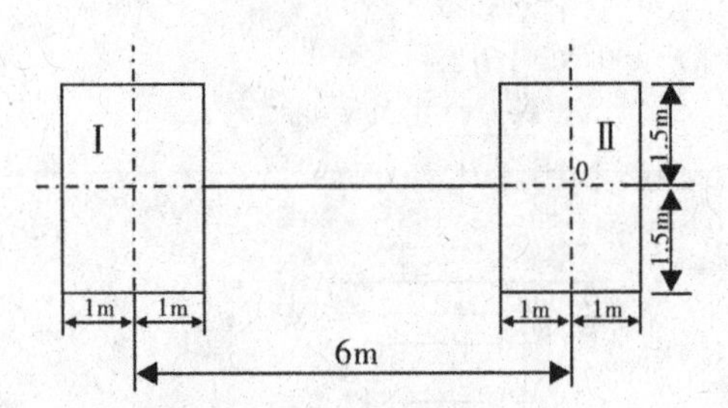


Fig. 3.32. Exercise 3.3.

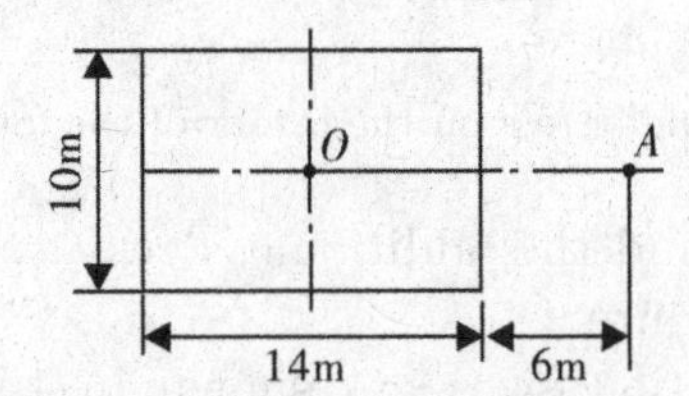


Fig. 3.33. Exercise 3.4.

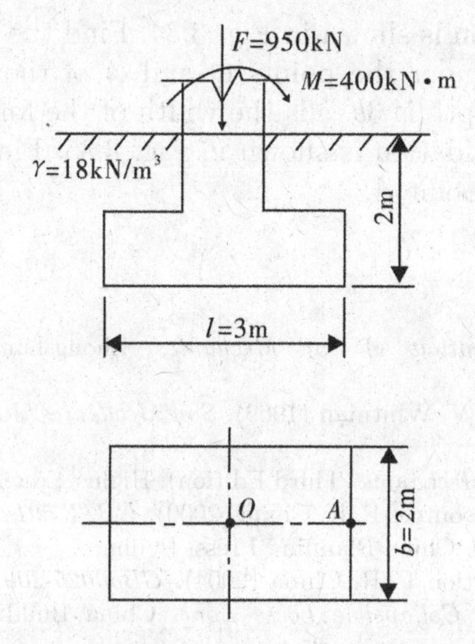


Fig. 3.34. Exercise 3.5.

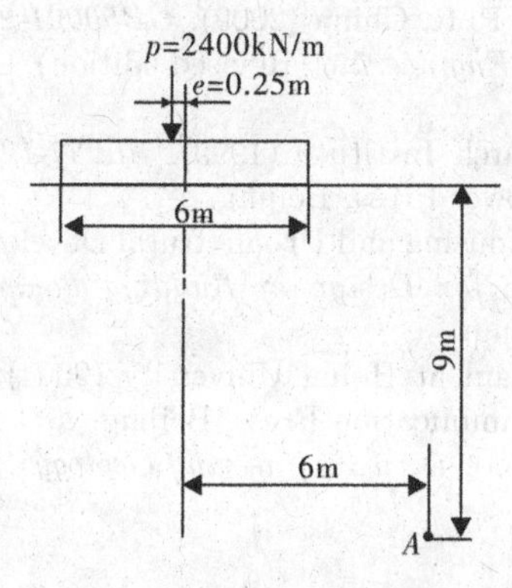


Fig. 3.35. Exercise 3.6.

additional stress distribution under the center of the foundation II, taking into account the influence of the adjoining foundation (the calculation depth is $z = 10\,\text{m}$, the unit weight of soil upward the floor is $18\,\text{kN/m}^3$).

3.4 The foundation is shown in Fig. 3.33. Find how many percentages of the additional stress at the point $A$ account for the point $B$ at the depth of 10 m.

3.5 The foundation is shown in Fig. 3.34. Find the additional stress distribution under the points $O$ and $A$ of the foundation (the calculation depth is $3b$, $b$ is the width of the foundation).
3.6 The strip foundation is shown in Fig. 3.35. Find the additional stress on the point $A$.

## Bibliography

S. He (2003). *Essentials of Soil Mechanics.* Zhongshan University Press, Guangzhou.

T. W. Lambe and R. V. Whitman (1969). *Soil Mechanics.* John Wiley and Sons, New York.

H. Liao (2018). *Soil Mechanics* (Third Edition). Higher Education Press, Beijing.

Ministry of Water Resources P. R. China (2000). *GB/T 50123-1999 Standard for Soil Test Method.* China Planning Press, Beijing.

Ministry of Construction P. R. China (2004). *GB50025-2007 Code for Building Construction in Collapsible Loess Zone.* China Building Industry Press, Beijing.

Ministry of Water Resources P. R. China (2008). *GB/T50145-2007 Standard for Engineering Classification of Soil.* China Planning Press, Beijing.

Ministry of Construction P. R. China (2009). *GB50021-2001 Code for Investigation of Geotechnical Engineering* (Revised edition). China Building Industry Press, Beijing.

Nanjing Hydraulic Research Institute (1999). *SL237-1999 Specification of Soil Test.* China Waterpower Press, Beijing.

P. R. China Ministry of Housing and Urban–Rural Development Producer (2009). *GB50007-2011 Code for Design of Building Foundation.* China Building Industry Press, Beijing.

Soil Mechanics Work Team at Hohai University (2004). *Soil Mechanics* (First edition). China Communication Press, Beijing.

S. Zhao and H. Liao (2009). *Civil Engineering Geology.* Science Press, Beijing.

# Chapter 4

# Compression and Consolidation of Soils

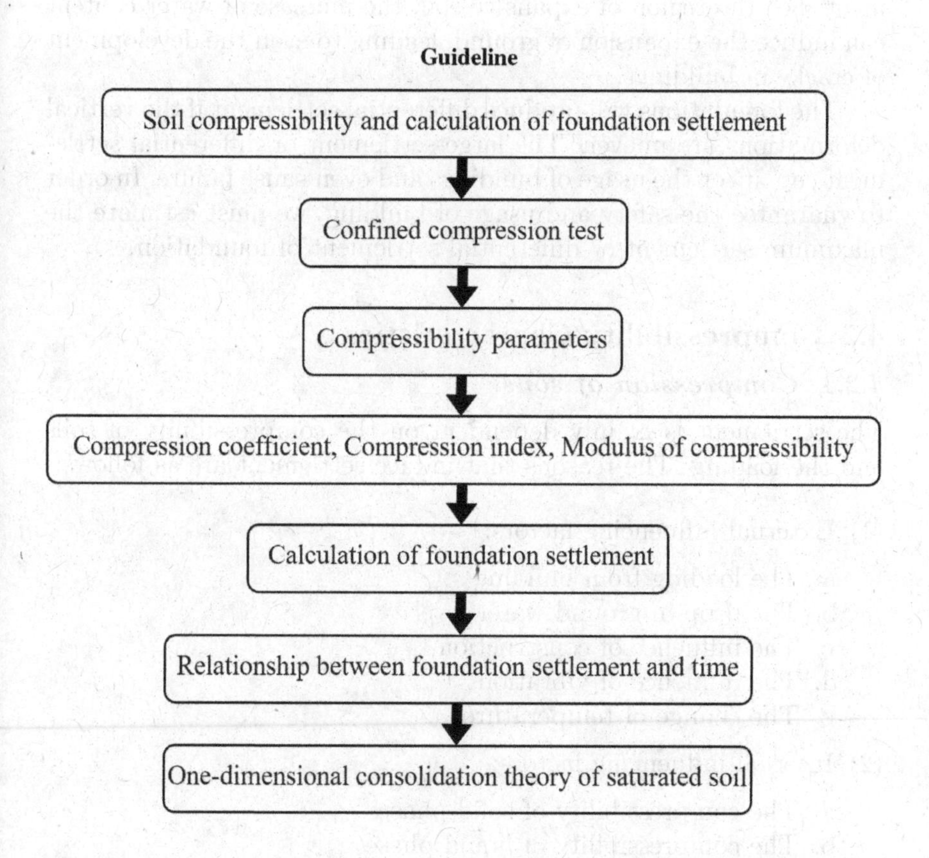

## 4.1 Introduction

There are two kinds of stress in soils, i.e., gravitational stress and additional stress. The deformation of soils is mainly caused by additional stress. Deformation occurs when the soil layers are subjected to the load from the buildings. The settlement is small as the soil is hard, which has no influence on the application of engineering methods. However, as the soil is soft and the thickness is uneven or as a result of the marked variance in the weight of the building, large settlement occurs, which will lead to accidents and affect the application of engineering methods.

For some special soils, deformation also occurs due to the change of water content. For collapsible loess, the additional settlement caused by the increase of water content is called collapsible settlement. For the region of expansive soil, the increase of water content can induce the expansion of ground, leading to even the development of cracks in buildings.

The foundations will produce differential settlement if the vertical deformations are uneven. The large settlement or differential settlement can affect the usage of buildings and even cause failure. In order to guarantee the safety and usage of building, we must estimate the maximum settlement or differential settlement of foundation.

## 4.2 Compressibility Characteristics

### 4.2.1 *Compression of soils*

The settlement is mainly dependent on the compressibility of soils and the loading. The reasons that induce settlement are as follows:

(1) External influencing factors:

a. The loading from building
b. The drop in ground water level
c. The influence of construction
d. The influence of vibrations
e. The change of temperature

(2) Internal influencing factors:

a. The compressibility of solid phase
b. The compressibility of liquid phase
c. The compressibility of void

As regards external factors, loading from buildings is the dominant factor, which causes compression of voids of soil. The process of the increase of the compression of saturated soil with time is called consolidation.

Settlement mainly includes two aspects: one is the total settlement; the other is the relationship of settlement and time. To determine the settlement, the compressibility index must be determined, which can be obtained from the compression test of the field *in situ* test.

### 4.2.2 *Oedometer test*

Figure 4.1 depicts a schematic diagram of how an oedometer test is conducted. Consolidation settlement is the vertical displacement of the surface corresponding to the volume change at any stage of the consolidation process. The characteristics of a soil during one-dimensional consolidation or swelling can be determined by means of the oedometer test. Figure 4.1 shows a cross-section diagram through an oedometer. The confining ring imposes a condition of zero lateral strain on the specimen.

Depending on the $\Delta H$–$p$ relationship, the $e$–$p$ curve is obtained, which indicates the change of void under varying pressure.

Figure 4.2(a) shows the $e$–$p$ relationship. It can also be plotted in a logarithmic scale (see Fig. 4.2(b)). For the different soil types, the deformations are different. For sands, the $e$–$p$ curve is flat, which indicates that the void ratio decreases slowly as pressure increases.

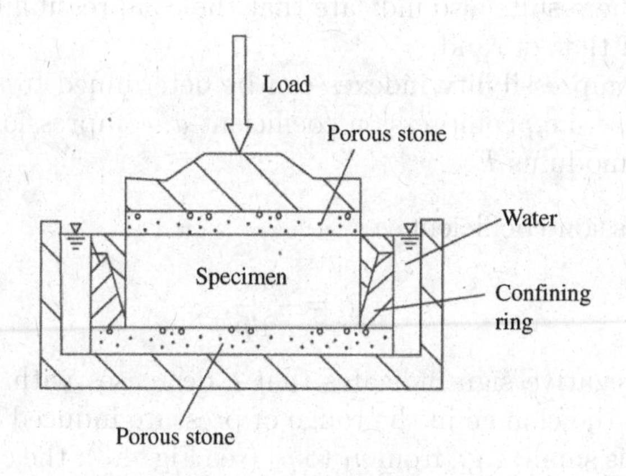


Fig. 4.1. Oedometer test.

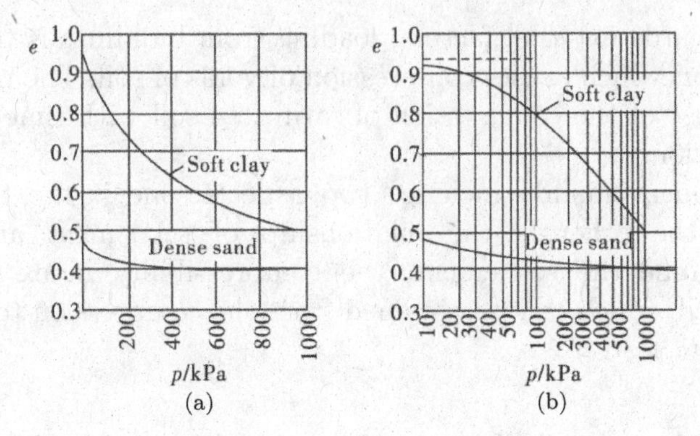


Fig. 4.2. Relationship between void ratio and effective stress. (a) $e$–$p$ and (b) $e$–lg$p$.

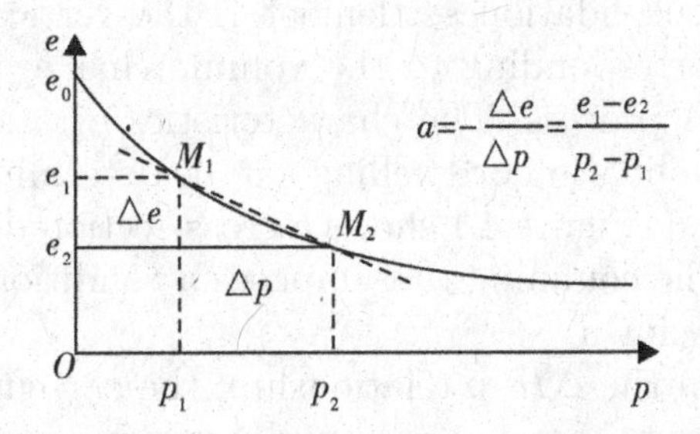


Fig. 4.3. The $e$–$p$ curve for soil.

For clay, the curve is steep, which means the void ratio decreases significantly as the pressure increases. For clay, the curve is steep, which means that the void ratio decreases significantly as pressure increases. The results also indicate that the compressibility of clay is greater than that of sand.

Three compressibility indexes can be determined from the compression curve, i.e., compression coefficient $a$, compression index $C_c$, oedometric modulus $E_s$.

(1) Compression coefficient, $a$

$$a = -\frac{de}{dp}, \tag{4.1}$$

where negative sign indicates that $e$ decreases with an increase in $p$. As the change in the range of pressure induced by external loading is small, e.g., from $p_1$ to $p_2$ (see Fig. 4.3), the curve $M_1M_2$

can be approximately regarded as linear. The slope of the line is give as follows:

$$a = \frac{\Delta e}{\Delta p} = \frac{e_1 - e_2}{p_2 - p_1}. \tag{4.2}$$

The compression coefficient $a$ represents the decreased value of void ratio under unit pressure. So the larger the value of $a$, the larger the compressibility of soils.

It should be noted that $a$ is not a constant for a certain soil type. In order to facilitate the comparison and the application of different regions, *Code for Design of Building Foundation (GB50007-2011), China*, takes $a_{1-2}$ corresponding to $p_1 =$ 100 kPa and to $p_2 = 200$ kPa to evaluate the compressibility:

$$\begin{aligned} a_{1-2} &< 0.1\ \text{MPa}^{-1} && \text{low compressibility soil,} \\ 0.1\ \text{MPa}^{-1} \leq a_{1-2} &< 0.5\ \text{MPa}^{-1} && \text{medium compressibility soil,} \\ a_{1-2} &\geq 0.5\ \text{MPa}^{-1} && \text{high compressibility soil.} \end{aligned}$$

(2) Compression index, $C_c$

The compression index ($C_c$) is the slope of the linear portion of the $e$–lg$p$ plot (see Fig. 4.4) and is dimensionless. For any two points on the linear portion of the plot, $C_c$ is given as follows:

$$Cc = \frac{e_1 - e_2}{\lg p_1 - \lg p_2} = \frac{e_1 - e_2}{\lg(\frac{p_2}{p_1})}. \tag{4.3}$$

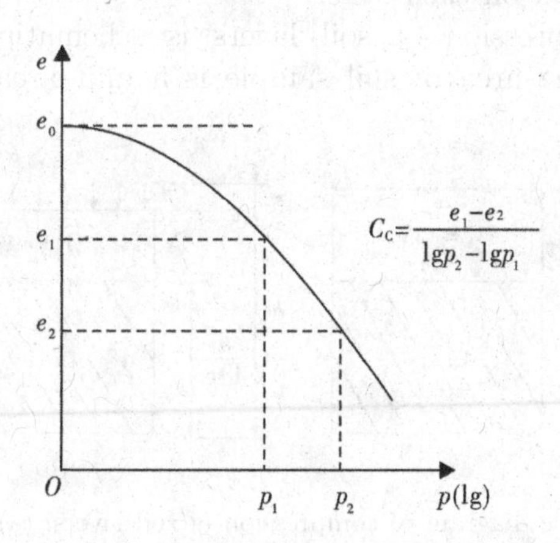


Fig. 4.4. Calculation of $C_c$ in $e$–lg$p$.

Similar to the compression coefficient, the value of $C_c$ can also judge the compressibility. The larger the value of $C_c$, the larger the void ratio is and the larger compressibility is. Generally, if $C_c < 0.2$, the soil belongs to the low compressibility soil type; if $C_c = 0.2$–$0.4$, the soil is the medium compressibility soil type; and if $C_c > 0.4$, soil is the high compressibility soil type.

(3) Oedometric modulus, $E_s$

The ratio of the incremental vertical stress $\sigma_z$ to stress $\lambda_z$ is called oedometric modulus $E_s$:

$$E_s = \frac{\sigma_z}{\lambda_z}. \tag{4.4}$$

In the above oedometer test, vertical pressure increases from $p_1$ to $p_2$, and the height decreases from $h_1$ to $h_2$, simultaneously.

$$\begin{aligned} &\text{Incremental stress:} \quad \sigma_z = p_2 - p_1. \\ &\text{Incremental strain:} \quad \lambda_z = \frac{h_1 - h_2}{h_1}. \\ &\text{Oedometric modulus:} \quad E_s = \frac{p_2 - p_1}{h_1 - h_2} h_1. \end{aligned} \tag{4.5}$$

(4) Relationship of oedometric modulus and compression coefficient

Both oedometric modulus and compression coefficient are commonly used to express the compressibility of ground in civil engineering. They are determined by the oedometer test. So they are dependent on each other.

The compression of soil layers is schematically shown in Fig. 4.5. The area of soil sample is a unit area. At the start

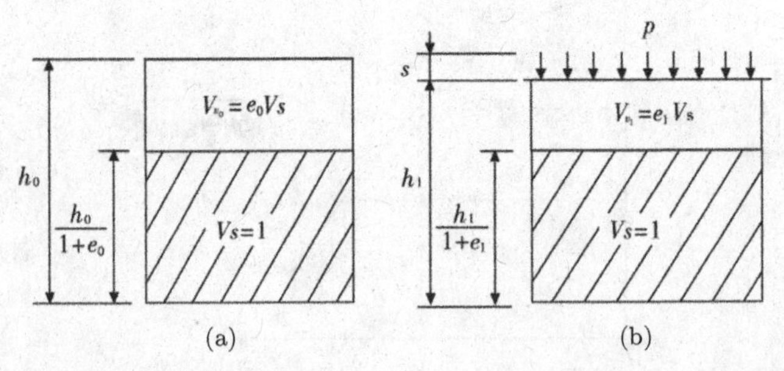


Fig. 4.5. Schematic diagram of compression of soil layers. (a) Initial state and (b) compression state.

of compression, the volume of solids is $V_s$, and the volume of voids is $V_{v0}$. Taking $V_s = 1$, the void ratio is $e_0 = V_{v0}$ and the total volume is $1 + e_0$ (see Fig. 4.5). At the end of compression, the volume of the solid $V_s$ remains constant and the volume of voids reduces to $V_{v1}$. The void ratio is $e_2 = V_{v2}$ (see Fig. 4.5). So

$$\lambda_z = \frac{V_{v0} - V_{v1}}{V_{v0}} = \frac{h_0 - h_1}{h_0} = \frac{e_0 - e_1}{1 + e_0}, \tag{4.6}$$

$$E_s = \frac{\sigma_z}{\lambda_z}\frac{p_2 - p_1}{e_0 - e_1}(1 + e_0) = \frac{1 + e_0}{a}. \tag{4.7}$$

It can be found from Eq. (4.7) that $E_s$ is inversely proportional to $a$. The larger $E_s$ is, the smaller $a$ is, which means the compressibility is smaller. In practice, if $E_s < 4$ MPa, the soil belongs to the high compressibility soil type, if 4 MPa $\leq E_s \leq 20$ MPa, the soil belongs to the medium compressibility soil type, and if $E_s > 20$ MPa, the soil belongs to the low compressibility soil type.

The compression of soil specimen under pressure is measured by means of a dial gauge operating on the loading cap. The void ratio ($e$) at the end of each increment period can be calculated from the dial gauge readings ($s$). The phase diagram is also shown in Fig. 4.5, and the calculation is done as follows:

$$h_0 A = V_{v0} + V_s = (1 + e_0)V_s, \tag{4.8}$$

$$(h_0 - s)A = V_{v1} + V_s = (1 + e_1)V_s, \tag{4.9}$$

$$V_s = \frac{h_0 A}{1 + e_0} = \frac{(h_0 - s)A}{1 + e_1}, \tag{4.10}$$

$$e_1 = e_0 - \frac{s}{h_0}(1 + e_0), \tag{4.11}$$

$$s = \frac{e_0 - e_1}{1 + e_0}h_0, \tag{4.12}$$

$$\varepsilon_v = \varepsilon_z = \frac{s}{h_0}$$

$$= \frac{e_{10} - e_1}{1 + e_0} = \frac{\Delta e}{1 + e_0}, \tag{4.13}$$

where $h_0$ is the thickness of the specimen at the start of the test and $e_0$ is the void ratio of the specimen at the start of the test, which can be determined as follows:

$$e_0 = \frac{d_s\,(1 + w_0)\,\rho_\omega}{\rho_0} - 1, \tag{4.14}$$

where $h_1$ and $e_1$ are the thickness and void ratio at the end of any increment period, respectively, $A$ is the area of the specimen, and $s$ is the compression of the specimen under pressure $p$.

The coefficient of volume compressibility $m_v$ is defined as the volume change per unit volume per unit increase in effective stress. If for an increase in effective stress from $\sigma_0\prime$ to $\sigma_1\prime$, the void ratio decreases from $e_0$ to $e_1$, then

$$m_v = \frac{\varepsilon_v}{\Delta\sigma'} = \frac{\Delta e}{(1 + e_0)\Delta\sigma'} = \frac{1}{1 + e_0}\left(\frac{e_0 - e_1}{\sigma_1' - \sigma_0'}\right). \tag{4.15}$$

The unit of $m_v$ is the inverse of pressure ($\text{m}^2/\text{MN}$).

If $m_v$ and $\Delta\sigma'$ are assumed constant with respect to depth, then the one-dimensional consolidation settlement ($s$) of the layer of thickness $h_0$ is given by

$$s = mv\Delta\sigma' h_0 \tag{4.16}$$

or, in the case of a normally consolidated clay,

$$s = \frac{C_c \lg(\sigma_1'/\sigma_0')}{1 + e_0} h_0. \tag{4.17}$$

## 4.3 Calculation of Settlement of Foundation

Settlement of buildings is induced by the deformation of soils. There are many methods to calculate the settlement of buildings, including the layerwise summation method, code method, and elastic method. Herein, only the layerwise summation method is presented.

For the calculation of the compression of soil, the layerwise summation method is most widely used in engineering, which is based on the formula of compression under the lateral confining condition. The assumptions are as follows:

(1) The compression of the soil is due to the deformation of the skeleton of the soil which is induced by the reduction of void volume. The soil particles are incompressible.

(2) Only vertical compression occurs in the soil, not lateral deformation.
(3) The pressure is distributed uniformly in the various soil layers.

### 4.3.1 *Calculation principle*

For the layer-wise summation method, the ground is first divided into several horizontal layers whose thicknesses are $h_1, h_2, h_3, \ldots, h_n$, respectively (see Fig. 4.6), and the compressions $s_1, s_2; s_3, \ldots, s_n$ of each soil layer are calculated, and then they all are summed together, which is the settlement of ground base,

$$s = s_1 + s_2 + \cdots + s_n = \sum_{i=1}^{n} s_i. \tag{4.18}$$

Calculation method and procedure for the layerwise summation method are as follows:

(1) Drawing the profile maps of ground and foundation according to the scale.
(2) Dividing the ground into several layers. A layer surface should be kept at the interface between the natural soil layer and the underground water level. The thickness of the soil layer for one type of soils should not be too large, that is, equal to or less than $0.4b$.

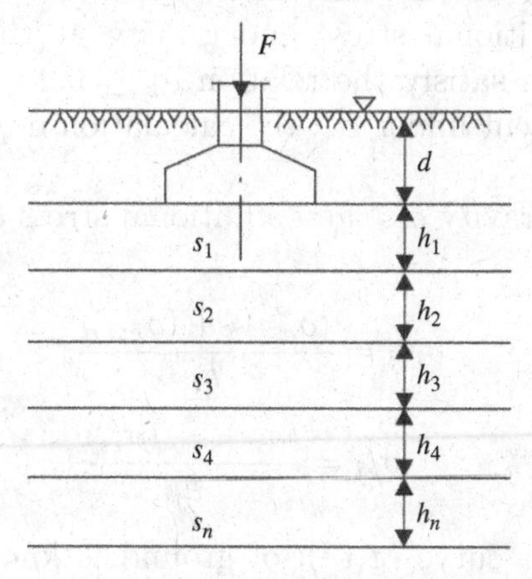


Fig. 4.6. Principle of the layerwise summation method.

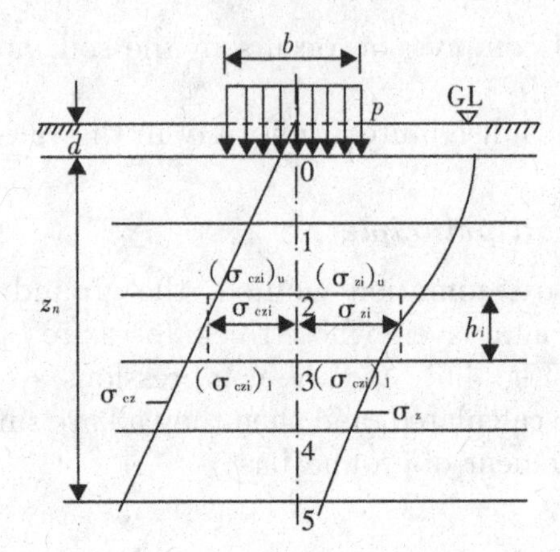


Fig. 4.7. Sample diagram showing one-dimensional method.

(3) The pressure in each layer is regarded as being uniformly distributed.
(4) The gravity stress and additional stress along the central axis of foundation and plot of the stress envelop are shown in Fig. 4.7. It should be noted that the base pressure $p = p_0 - \gamma d$ ($\gamma$ is the weight of the soil in the range of embedded depth of foundation) is used to calculate the additional pressure in the ground (from the base of the foundation).
(5) Determining the calculation depth of settlement. Previous experiences show that the depth is called the lower limit of compression layers or the calculation depth of the settlement $Z_n$ when the additional stress and gravity at the central axis of the foundation satisfy the relation $\sigma_z \leq 0.2\sigma_{cz}$. When soft soil layers are present under $Z_n$, the calculation depth should satisfy $\sigma_z \leq 0.1\sigma_{cz}$.
(6) The average gravity $\sigma_{czi}$ and additional stress $\sigma_{zi}$ are calculated as follows:

$$\sigma_{czi} = \frac{(\sigma_{czi})_u + (\sigma_{czi})_l}{h_i}, \tag{4.19}$$

$$\sigma_{zi} = \frac{(\sigma_{zi})_u + (\sigma_{zi})_l}{h_i}. \tag{4.20}$$

(7) If compression curve of $e$–$p$ of ground is known, according to the initial stress and additional stress of the $i$-th layer, i.e.,

$p_{2i} = \sigma_{czi} + \sigma_{zi}$, the corresponding initial void ratio and the stable void ratio can be obtained based on the compression curve.

(8) According to $s = \frac{e_1 - e_{21}}{1+e_1}h$, the compression of the $i$-th layer can be calculated, i.e.,

$$s = \frac{e_{1i} - e_{2i}}{1+e_{1i}} h_i.$$

(9) Finally, the settlement of foundation is the summation of compression of all layers,

$$s = \sum_{i=1}^{n} s_i.$$

Sometimes, if by using the exploration unit we are not able to obtain the compression curve, but are able to determine the other compression indicators, such as compression coefficient and confined compression modulus, then the formulations $s_i = \frac{a_i}{1+e_{1i}}\overline{\sigma_{zi}}h_i$ and $s_i = \frac{\overline{\sigma_{zi}}}{E_{si}}h_i$ are used to calculate the compression of each soil layer. The total settlement of foundation can be obtained by summing the compressions of all layers.

The shortcomings of the layerwise summation method can be analyzed from the following factors: the calculation and distribution of additional stress, the selection of compression indicators, and the thickness of the soil layers. The method adopts the elastic theory to calculate the vertical stress $\sigma_z$ and uses the $e$–$p$ curve to determine the uniaxial compression and deformation, which deviates from the stress condition of the ground base.

For the deformation index, the test condition determines its results, and the selection method also affects the calculation results. Furthermore, for measuring the thickness of the compression layer, there is no strict theoretical basis. The determination of the thickness is based on a semi-empirical method, which is only verified from engineering measurement. The above factors result in the fact that the different methods that determine the thickness of the compressed layer can produce an error of 10%. However, the concept of the layer-wise summation method of settlement calculation is clear. The calculation process and deformation index are relatively simple and easy to be grasped. So it is still widely used in engineering.

## 4.4 One-Dimensional Consolidation Theory

The mechanics of the one-dimensional consolidation process can be represented by means of a simple analogy as shown in Fig. 4.8.

Figure 4.8(a) shows a spring inside a cylinder filled with water and a piston, on top of the spring. It is assumed that there can be no leakage and no friction between the piston and the cylinder. The spring represents the compressible soil skeleton, the water in the cylinder represents the pore water, and the bore diameter of the valve is regarded as the permeability of the soil. The cylinder itself simulates the condition of no lateral strain in the soil. Suppose a load is now placed on the piston with no leakage, as in Fig. 4.8(b). Assuming water is incompressible, the piston will not move, with the result that no load can be transmitted to the spring; the load will be carried by the water, with the increase in pressure in the water being equal to the load divided by the piston area. This situation corresponds to the undrained condition in the soil. If the drain holes are now opened, water will be forced out through the drain holes at a rate governed by the bore diameter. This will allow the piston to move and the spring to be compressed as load is gradually transferred to it. This situation is shown in Fig. 4.8(c). At any time an increase in the load on the spring will correspond to a reduction in pressure in the water. Eventually, as shown in Fig. 4.8(d), all the load will be carried by the spring and the piston will come to rest. This corresponds to the drained condition in the soil. At any time, the load carried by the spring represents the effective normal stress in the soil, the pressure of the water in the cylinder the pore water pressure and the load on the piston the total normal stress. The movement of the piston

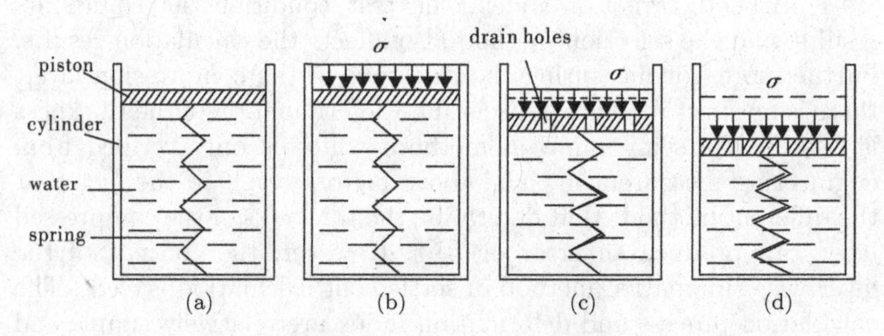


Fig. 4.8. Consolidation analogy.

represents the change in volume of the soil and is governed by the compressibility of the spring (the equivalent of the compressibility of the soil skeleton). The piston and spring analogy represents only an element of soil since the stress conditions vary from point to point throughout a soil mass.

### 4.4.1 *Terzaghi's theory of one-dimensional consolidation*

The assumptions made in the theory of one-dimensional consolidation are as follows:

(1) The soil is homogeneous and fully saturated.
(2) The solid particles and water are incompressible.
(3) Compression and flow are one dimensional (vertical).
(4) Strains are small.
(5) Darcy's law is valid at all hydraulic gradients.
(6) The coefficient of permeability and the coefficient of volume compressibility remain constant throughout the process.
(7) There is a unique relationship, independent of time, between the void ratio and effective stress.

Consider an element of fully saturated soil having dimensions $dx$, $dy$, and $dz$ in the $x$, $y$, and $z$ directions, respectively, within a clay layer of thickness $H$, as shown in Fig. 4.9. An increment of total vertical stress $p_0$ is applied to the element with flow taking place in the $z$ direction only. The component of discharge velocity of

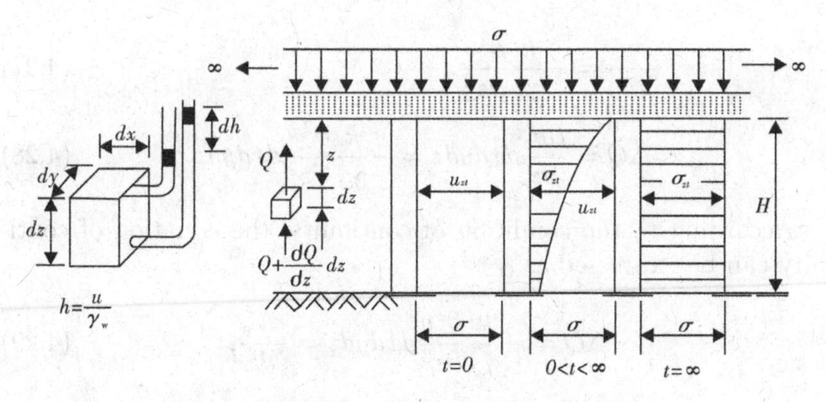


Fig. 4.9. Seepage through a soil element.

water entering the element is $v_z$, and the rate of change of discharge velocity in the $z$ direction is $\partial v_z/\partial z$. The volume of water entering the element per unit time is

$$Q_e = v_z dxdy. \tag{4.21}$$

And the volume of water leaving per unit time is

$$Q_l = v_z dxdy + \frac{\partial v_z dxdy}{\partial z} dz. \tag{4.22}$$

If water is assumed to be incompressible, then the difference between the volume of water entering the element per unit time and the volume leaving is

$$\Delta Q = Ql - Qe = \frac{\partial v}{\partial z} dxdydz. \tag{4.23}$$

The flow velocity through the element is given by Darcy's law as

$$vz = ki = -k\frac{\partial h}{\partial z}. \tag{4.24}$$

Since any change in total head ($h$) is due only to a change in pore water pressure, we have

$$u = \gamma_\omega h, \quad h = \frac{u}{\gamma_\omega}, \tag{4.25}$$

$$vz = ki = -k\frac{\partial h}{\partial z} = -\frac{k}{\gamma_\omega}\frac{\partial u}{\partial z}, \tag{4.26}$$

$$\frac{\partial v_z}{\partial z} = -\frac{k}{\gamma_\omega}\frac{\partial^2 u}{\partial z^2}, \tag{4.27}$$

$$\Delta Q = \frac{\partial v_z}{\partial z} dxdydz = -\frac{k}{\gamma_\omega}\frac{\partial^2 u}{\partial z^2} dxdydz. \tag{4.28}$$

According to the condition of continuity, the equation of continuity can be expressed as

$$\Delta Q = -\frac{k}{\gamma_\omega}\frac{\partial^2 u}{\partial z^2} dxdydz = \frac{dV_v}{dt}, \tag{4.29}$$

where $dV_v/dt$ is the volume change per unit time.

The rate of volume change can be expressed as

$$\frac{dV_v}{dt} = \frac{\partial}{\partial t}\left(\frac{e}{1+e_0}dxdydz\right) = \frac{1}{1+e_0}\frac{\partial e}{\partial t}dxdydz = m_v\frac{\partial \sigma'}{\partial t}dxdydz. \tag{4.30}$$

The total stress increment is gradually transferred to the soil skeleton, increasing the effective stress, as the excess pore water pressure decreases. Hence, the rate of volume change can be expressed as

$$\partial\sigma' = \partial(\sigma - u) = -\partial u, \tag{4.31}$$

$$\frac{dV_v}{dt} = m_v\frac{\partial u}{\partial t}dxdydz. \tag{4.32}$$

Combining Eqs. (4.29) and (4.32), we have

$$\frac{k^2 u}{\gamma_w z^2}dxdydz = mv\frac{u}{t}dxdydz \tag{4.33}$$

or

$$\frac{\partial u}{\partial t} = \frac{k}{m_v\gamma_w}\frac{\partial^2 u}{\partial z^2} = C_v\frac{\partial^2 u}{\partial z^2}. \tag{4.34}$$

This is the differential equation of consolidation, where $C_v$ is defined as the coefficient of consolidation, with the suitable unit being $m^2$/year. Since $k$ and $m_v$ are assumed as constants, $C_v$ is constant during consolidation.

### 4.4.2 *Solution of the consolidation equation*

Consider the initial and boundary conditions of a clay layer of thickness $2H$, as shown in Fig. 4.10, the solution for the differential equation of consolidation is as follows:

The total stress increment is assumed to be applied instantaneously. At zero time, therefore, the increment will be carried entirely by the pore water, i.e., the initial value of excess pore water pressure ($u$) is

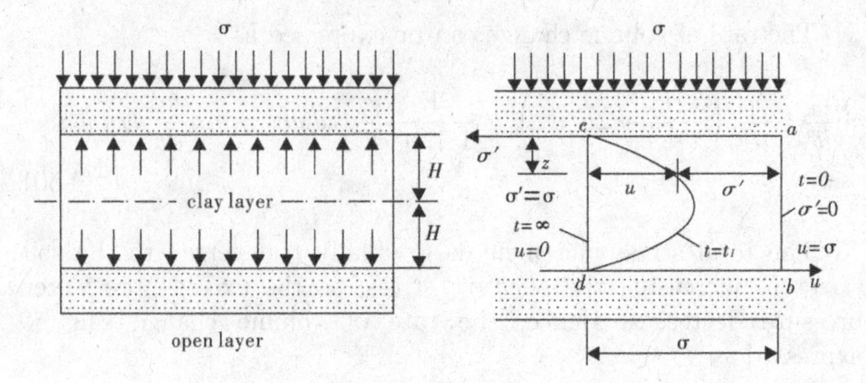


Fig. 4.10. Distribution of excess pore water pressure.

equal to $\sigma$ and the initial condition is

$$u = \sigma, \quad \text{for } 0 \leq z \leq 2H \text{ when } t = 0. \tag{4.35}$$

The upper and lower boundaries of the clay layer are assumed to be free-draining, with the permeability of the soil adjacent to each boundary being very high compared to that of the clay. Thus, the boundary conditions at any time after the application of $\sigma$ are

$$u = 0,\ \sigma' = \sigma, \quad \text{for } z = 0 \text{ and } z = 2H \text{ when } 0 < t < \infty, \tag{4.36}$$

$$u = 0, \sigma' = \sigma, \quad \text{for } 0 \leq z \leq 2H \text{ when } t = \infty. \tag{4.37}$$

The solution for the excess pore water pressure at depth $z$ after time $t$ is

$$u_{zt} = \frac{4\sigma}{\pi} \sum_{m=1}^{\infty} \frac{1}{m} e^{-\frac{m^2\pi^2}{4}T_v} \sin \frac{m\pi z}{2H}, \tag{4.38}$$

or

$$u_{zt} = 2\sigma \sum_{m=1}^{\infty} \frac{1}{M} e^{-M^2 T_v} \sin \frac{Mz}{H}, M = \frac{m\pi}{2}, \tag{4.39}$$

where $m$ are odd values. $T_v$ is a dimensionless number called the time factor and is expressed as

$$T_v = \frac{C_v t}{H^2}. \tag{4.40}$$

The progress of consolidation can be shown by plotting a series of curves of $u$ against $z$ for different values of $t$. Such curves are called isochrones and their form will depend on the initial distribution of excess pore water pressure and the drainage conditions at the boundaries of the clay layer.

The degree of consolidation can be expressed as

$$u_{zt} = \frac{\sigma'_{zt}}{\sigma} = \frac{\sigma - u_{zt}}{\sigma} = 1 - \frac{u_{zt}}{\sigma}. \tag{4.41}$$

Combining Eqs. (4.38) and (4.41) or combining Eqs. (4.39) and (4.41), we have

$$U_{zt} = 1 - \frac{u_{zt}}{\sigma} = 1 - \frac{4}{\pi}\sum_{m=1}^{\infty}\frac{1}{m}e^{-\frac{m^2\pi^2}{4}T_v}\sin\frac{m\pi z}{2H} \tag{4.42}$$

or

$$U_{zt} = 1 - \frac{u_{zt}}{\sigma} = 1 - \sum_{m=1}^{\infty}\frac{2}{M}e^{-M^2T_v}\sin\frac{Mz}{H}, M = \frac{m\pi}{2}. \tag{4.43}$$

In practical problems, it is the average degree of consolidation ($U_t$) over the depth of the layer as a whole that is of interest, with the consolidation settlement at time $t$ ($S_t$) being given by the product of $U_t$ and the final settlement ($S$). The average degree of consolidation at time $t$ for constant $\sigma$ is given by

$$U_t = \frac{S_t}{S} = \frac{m_v\int_0^{2H}\sigma' dz}{m_v\int_0^{2H}\sigma dz} = \frac{\int_0^{2H}(\sigma - u_{zt})dz}{\int_0^{2H}\sigma dz} = 1 - \frac{\int_o^{2H}u_{zt}dz}{2H\sigma} \tag{4.44}$$

Combining Eqs. (4.39) and (4.44), we have

$$U_t = 1 - \sum_{m=1}^{\infty}\frac{2}{M^2}e^{-M^2T_v} \tag{4.45}$$

or

$$U_t = 1 - \frac{8}{\pi^2}e^{-M^2T_v} = 1 - \frac{8}{\pi^2}e^{-\frac{\pi^2}{4}T_v} \text{ when } m = 1. \tag{4.46}$$

The relationship between $U_t$ and $T_v$ is also represented by curves, as shown in Fig. 4.11, where the parameter $\alpha$ equals $\sigma'_z/\sigma''_z$, where $\sigma'_z$

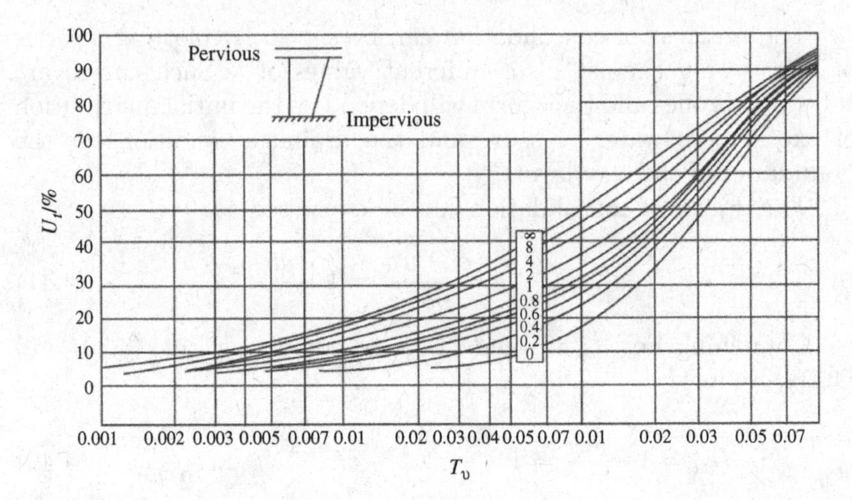


Fig. 4.11. Relationship of average degree of consolidation and time factor.

is the consolidation pressure on the pervious boundary surface and $\sigma_z''$ is the consolidation pressure on the impervious boundary surface. For the case of two-way drainage, i.e., open layer, the average degree of consolidation can also be calculated using Fig. 4.11, but $\alpha$ is equal to 1 and the drainage path $H$ should be half the thickness of the layer.

According to Eq. (4.46), with the distribution of consolidation pressure and as a result of the drainage condition, the following two types of engineering problems can be resolved:

(1) With the final settlement $S$ known, find the settlement $S_t$ at an elapsed consolidation time $t$.

For such a problem, first the average coefficient of consolidation $Cv$ and the time factor $T_v$ of the soil layer can be calculated using the known values of $k, a, e_1, H$, and $t$. Then the corresponding average degree of consolidation $U_t$ can be read off from Fig. 4.11. Therefore, $S_t$ can be calculated.

(2) With the final settlement $S$ known, find the length of time $t$ for achieving a settlement $S_t$.

For such a problem, first the average degree of consolidation $U_t$ is $S_t/S$, then the time factor $T_v$ of the soil layer can be read off from Fig. 4.11. Therefore, the length of time $t$ required can be calculated using Eq. (4.46).

## Exercises

4.1 An undisturbed saturated sample was cut with a ring cutter with an inner diameter of 7.98 cm and a height of 2 cm. The relative density is 2.70, the moisture content is 36.3%, and the wet weight is 184 g. The oedometer test was carried out. Under the pressure of 100 and 200 kN/m$^2$, the compressions of the sample were 1.0 mm and 1.6 mm, respectively. Please calculate the void ratio after compression, the compression coefficient, and the compression modulus of the soil.

4.2 There is a sedimentary layer under the water surface. A sample is taken from the ground below 5 m depth. The moisture content is 180%, the gravity is 12.8 kN/m$^3$, and the relative density of soil particles is 2.59. From the $e$–lg$p$ curve of the soil sample, the preconsolidation pressure is determined as 12.0 kN/m$^2$. Please try to determine the consolidation state of the soil layer.

4.3 The base area of a hotel column foundation is 4.00 m $\times$ 4.00 m, and the buried depth of the foundation is $d = 2.00$ m. The center load transmitted from the upper structure to the top of the foundation is 4720 kN. The surface layer of the foundation is fine sand with the gravity $\gamma_1 = 17.5\,\text{kN/m}^3$, the oedometric modulus $E_{s1} = 8\,\text{MPa}$, and a thickness $h_1 = 6.00$ m. The second layer is silty clay with $E_{s2} = 3.33\,\text{MPa}$ and $h_2 = 3.00$ m. The third layer is gravel with $E_{s3} = 22\,\text{MPa}$ and $h_3 = 4.50$ m. We suggest readers use the method of layered summation to calculate the settlement of silty clay layer.

4.4 The thickness of a clay layer is 4 m and it is a half-closed layer. The infinity uniform load is applied on the surface and the final settlement is 28 cm. After 100 days, the settlement of the clay layer is 18.5 cm and the relationship between the degree of consolidation and time factor is $U = 1.128(T_v)^{\frac{1}{2}}$. Try to find the coefficient of consolidation $C_v$.

4.5 The width of a rectangular foundation is 4 m. The additional stress on the floor is 100 kPa. Its embedment depth is 2 m. There are two soil layers in the depth scope 12 m. The thickness of the upper soil is 6 m and its bulk unit weight is 18 kN/m$^3$ and the relationship between its void ratio and compression pressure is $e = 0.85 - \frac{2}{3}p$. The thickness of the subsoil is 6 m and its bulk unit weight is 20 kN/m$^3$, and the relationship between

Table 4.1. The additional and the average additional stress coefficients.

| Depth (m) | 0 | 1 | 2 | 3 | 4 | 5 |
|---|---|---|---|---|---|---|
| The additional stress coefficient | 1.0 | 0.94 | 0.75 | 0.54 | 0.39 | 0.28 |
| The average additional stress coefficient | 1.0 | 0.98 | 0.92 | 0.83 | 0.73 | 0.65 |
| **Depth (m)** | **6** | **7** | **8** | **9** | **10** | |
| The additional stress coefficient | 0.21 | 0.17 | 0.13 | 0.11 | 0.09 | |
| The average additional stress coefficient | 0.59 | 0.53 | 0.48 | 0.44 | 0.41 | |

its void ratio and compression pressure is $e = 1.0 - p$. The water table is 6 m. The additional stress coefficient and the average additional stress coefficient distribution under the center of the foundation are shown in Table 4.1. Find the foundation settlement using the layerwise summation method and the recommendation settlement method by code (the coefficient of settlement is 1.05).

# Bibliography

S. He (2003). *Essentials of Soil Mechanics.* Zhongshan University Press, Guangzhou.

H. Liao (2018). *Soil Mechanics* (Third Edition). Higher Education Press, Beijing.

Ministry of Construction P. R. China (2009). *GB50021-2001 Code for Investigation of Geotechnical Engineering* (2009 edition). China Building Industry Press, Beijing.

Ministry of Construction P. R. China (2019). *GB50025-2018 Code for Building Construction in Collapsible Loess Zone.* China Building Industry Press, Beijing.

Ministry of Water Resources P. R. China (2000). *GB/T 50123-1999 Standard for Soil Test Method.* China Planning Press, Beijing.

Ministry of Water Resources P. R. China (2008). *GB/T50145-2007 Standard for Engineering Classification of Soil.* China Planning Press, Beijing.

Nanjing Hydraulic Research Institute (1999). *SL237-1999 Specification of Soil Test.* China Waterpower Press, Beijing.

P. R. China Ministry of Housing and Urban–Rural Development Producer (2009). *GB50007 2011 Code for Design of Building Foundation.* China Building Industry Press, Beijing.

Soil Mechanics Work Team at Hohai University (2004). *Soil Mechanics.* China Communication Press, Beijing.

S. Zhao and H. Liao (2009). *Civil Engineering Geology.* Science Press, Beijing.

# Chapter 5

# Shear Strength

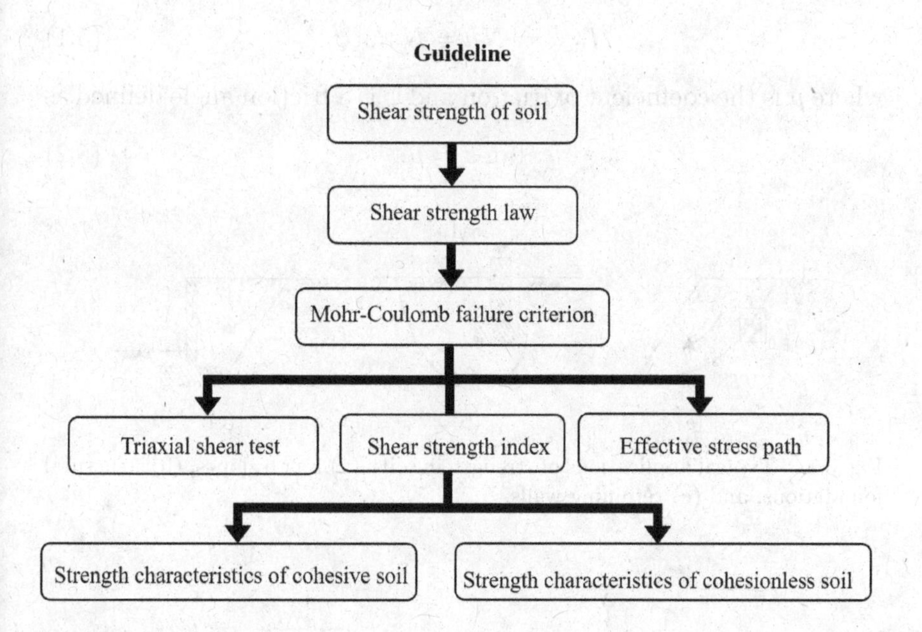

## 5.1 Shear Resistance

Soil is a granular material, so shear failure occurs when stresses between the particles are such that they slide or roll each other. Hence, the shear strength of soil is mainly controlled by friction. If at a point on any plane within a soil mass the shear stress becomes equal to the shear strength of the soil, the failure will occur at that point, as shown in Fig. 5.1.

The shear resistance between two particles is the force that must be applied to cause a relative movement between the particles. If $N$ is the normal force across a surface, the maximum shear force on this surface is proportional to the normal force, as shown in Fig. 5.2(a), i.e.

$$T_{\max} = N\mu = N \tan \varphi, \tag{5.1}$$

where $\mu$ is the coefficient of friction and $\varphi$ is a friction angle defined as

$$\tan \varphi = \mu. \tag{5.2}$$

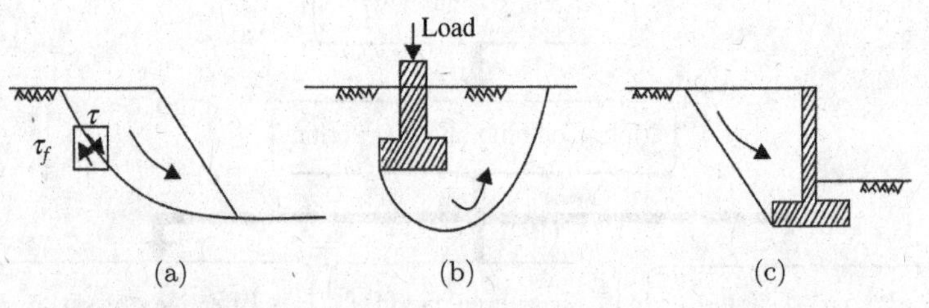


Fig. 5.1. Typical applications of strength in soils. (a) Earth slopes, (b) structural foundations, and (c) retaining walls.

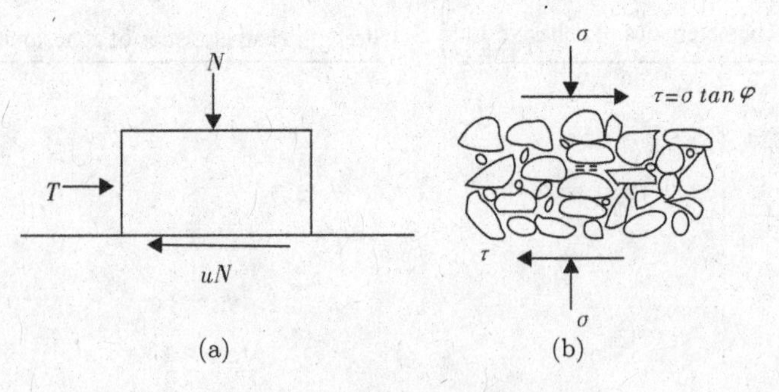


Fig. 5.2. Shear resistance.

The resistance of soil to deformation is strongly influenced by the shear resistance between particles. It is generally assumed that the relation between the normal stress $\sigma$ on every section through a mass of cohesive soil and the corresponding shearing resistance $\tau$ per unit of area can be represented by an empirical equation

$$\tau = c + \sigma \tan \varphi, \tag{5.3}$$

where $c$ represents the cohesion, which is equal to the shearing resistance per unit of area if $\sigma = 0$. The equation is known as Coulomb's equation. For cohesionless soil ($c = 0$) as shown in Fig. 5.2(b), the corresponding equation is

$$\tau = \sigma \tan \varphi. \tag{5.4}$$

The values $c$ and $\varphi$ contained in the preceding equations can be determined by means of laboratory tests, by measuring the shearing resistance on plane sections through the soil at different values of the normal stress $\sigma$. States of stress in two dimensions can be represented on a plot of shear stress ($\tau$) against normal stress ($\sigma$). The relation between $\sigma$ and $\tau$ is shown in Fig. 5.3. The shear strength ($\tau_f$) of soil at a point on a particular plane is also expressed by Coulomb as a linear function of the normal stress at failure ($\sigma_f$) on the plane at the same point.

$$\tau_f = c + \sigma_f \tan \varphi \text{ for cohesive soil} \tag{5.5}$$

$$\tau_f = \sigma_f \tan \varphi \text{ for cohesionless soil} \tag{5.6}$$

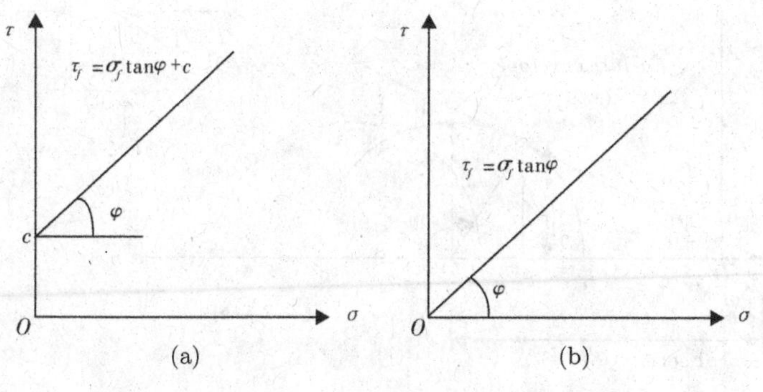


Fig. 5.3. Relation between $\sigma$ and $\tau$. (a) Cohesive soil and (b) cohesionless soil.

The $c$ and $\varphi$ are also called the shear strength parameters referred to as the cohesion intercept and the angle of shearing resistance, respectively.

However, in accordance with the principle of effective stress, the shear stress in a soil can be resisted only by the skeleton of solid particles. The shear strength should be expressed as a function of the effective normal stress at failure ($\sigma'_f$), and the effective shear strength parameters are denoted $c'$ and $\varphi'$ and need to be determined by either laboratory or *in situ* tests.

$$\tau_f = c' + \sigma'_f \tan\varphi' \text{ for cohesive soil,} \tag{5.7}$$

$$\tau_f = \sigma'_f \tan\varphi' \text{ for cohesionless soil.} \tag{5.8}$$

## 5.2 Mohr–Coulomb Failure Criterion

A stress state can be represented either by a point with coordinates $\tau$ and $\sigma(\sigma')$ as shown in Fig. 5.4 or by a Mohr circle defined by the effective principal stresses $\sigma_1(\sigma'_1)$ and $\sigma_3(\sigma'_3)$. Mohr circles representing stress states at failure are shown in Fig. 5.4, the compressive stress being taken as positive. The line touching the Mohr circles may be straight or slightly curved and is referred to as the failure envelope. A state of stress represented by a Mohr

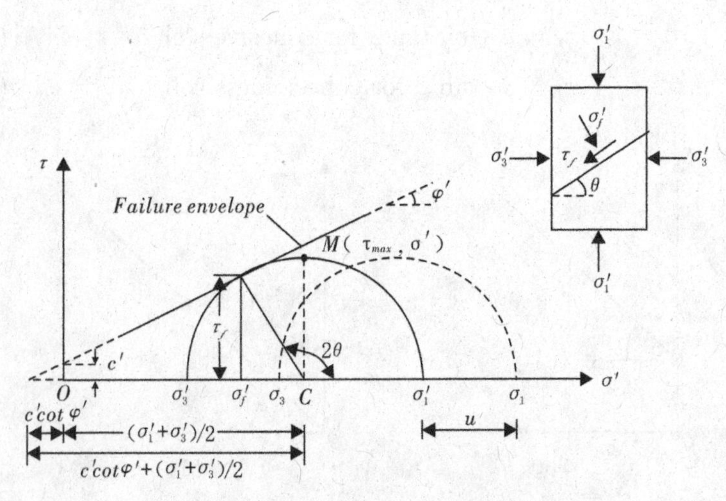


Fig. 5.4. Mohr–Coulomb failure criterion.

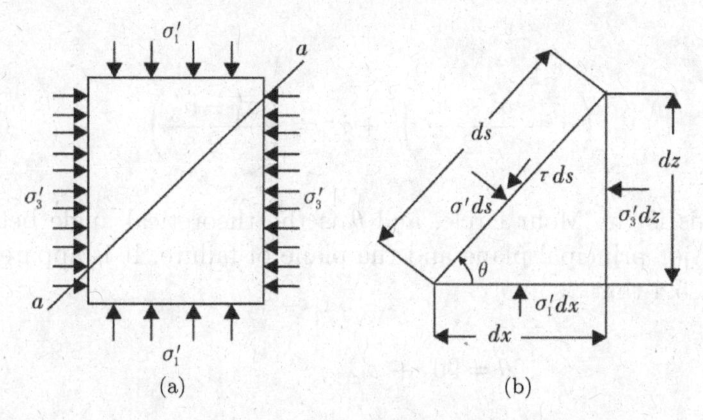


Fig. 5.5. Stress conditions in soil during triaxial compression test.

circle, part of which lies above the envelope, is impossible. For a given state of stress, it is apparent that, because $\sigma_1' = \sigma_1 - u$ and $\sigma_3' = \sigma_3 - u$, the Mohr circles for total and effective stresses have the same diameter but their centers are separated by the corresponding pore water pressure $u$. Similarly, the total and effective stress points are separated by the value of $u$.

In order to determine the stresses on an arbitrarily inclined section a–a through the specimen shown in Fig. 5.5(a), we investigate the conditions for the equilibrium of a small prism shown in Fig. 5.5(b), one side of which is located on the inclined section. The other two sides are parallel to the direction of the principal stresses $\sigma_1$ and $\sigma_3$, respectively. The slope of the inclined surface is determined by the angle $\theta$. The equilibrium of the prism requires that

$$\sum F_x = \sigma' \sin\theta ds - \sigma_3' \sin\theta - \tau \cos\theta ds = 0, \tag{5.9}$$

$$\sum F_z = \sigma_1' \cos\theta ds - \sigma' \cos\theta - \tau \sin\theta ds = 0. \tag{5.10}$$

Solving the above equations, the values of the stresses $\sigma$ and $\tau$ on the slope surface can be computed by the numerical values for $\sigma_1$, $\sigma_3$ and $\theta$:

$$\sigma' = \frac{\sigma_1' + \sigma_3'}{2} + \frac{\sigma_1' - \sigma_3'}{2} \cos 2\theta, \tag{5.11}$$

$$\tau = \frac{\sigma_1' - \sigma_3'}{2} \sin 2\theta, \tag{5.12}$$

or

$$\left(\sigma' - \frac{\sigma_1' + \sigma_3'}{2}\right)^2 + \tau^2 = \left(\frac{\sigma_1' - \sigma_3'}{2}\right)^2. \tag{5.13}$$

This is the Mohr circle, and $\theta$ is the theoretical angle between the major principal plane and the plane of failure. It is apparent as in Fig. 5.4 that

$$2\theta = 90^\circ + \varphi', \tag{5.14}$$

$$\theta = 45^\circ + \frac{\varphi'}{2}, \tag{5.15}$$

$$\sin\varphi' = \frac{\frac{1}{2}(\sigma_1' - \sigma_3')}{c\cot\varphi' + \frac{\sigma_1'+\sigma_3'}{2}} \Rightarrow \sin\varphi' = \frac{\sigma_1' - \sigma_3'}{2c\cot\varphi' + \sigma_1' + \sigma_3'}. \tag{5.16}$$

Therefore,

$$\sigma_1' = \sigma_3' \tan^2\left(45^\circ + \frac{\varphi'}{2}\right) + 2c\tan\left(45^\circ + \frac{\varphi'}{2}\right) = \sigma_3' \frac{1+\sin\varphi'}{1-\sin\varphi'} + 2c\frac{\cos\varphi'}{1-\sin\varphi'}, \tag{5.17}$$

$$\sigma_3' = \sigma_1' \tan^2\left(45^\circ - \frac{\varphi'}{2}\right) - 2c\tan\left(45^\circ - \frac{\varphi'}{2}\right) = \sigma_1' \frac{1-\sin\varphi'}{1+\sin\varphi'} - 2c\frac{\cos\varphi'}{1+\sin\varphi'}. \tag{5.18}$$

These are referred to as the Mohr–Coulomb failure criterion.

The state of stress represented in Fig. 5.4 could also be defined by the coordinates of point $M$, rather than by the Mohr circle. The coordinates of $M$ are $(\sigma_1' - \sigma_3')/2$ and $(\sigma_1' + \sigma_3')/2$, also denoted by $\tau_{\max}$ and $\sigma'$, which denote the maximum shear stress and the average principal stress, respectively. The stress state could also be expressed in terms of total stress. It should be noted that

$$\frac{\sigma_1' - \sigma_3'}{2} = \frac{\sigma_1 - \sigma_3}{2}, \tag{5.19}$$

$$\frac{\sigma_1' + \sigma_3'}{2} = \frac{\sigma_1 + \sigma_3}{2} - u. \tag{5.20}$$

Stress point $M$ lies on a modified failure envelope as shown in Fig. 5.6 defined by the equation

$$q = a' + p' \tan \alpha', \tag{5.21}$$

where $a'$ and $\alpha'$ are the modified shear strength parameters, and the relations among the parameters $a'$, $\alpha'$, $c'$ and $\varphi'$ are given approximately in Fig. 5.6.

$$\tan \alpha' = \sin \varphi' = \frac{R}{O'A} \Rightarrow \alpha' = \tan^{-1}(\sin \varphi'), \tag{5.22}$$

$$\tan \varphi' = \frac{c'}{O'O}, \tag{5.23}$$

$$\tan \alpha' = \frac{a'}{O'O} = \frac{a'}{c'} \tan \varphi' = \frac{a'}{c'} \frac{\sin \varphi'}{\cos \varphi'}. \tag{5.24}$$

Combining Eqs. (5.22) and (5.24), we have

$$a' = c' \cos \varphi'. \tag{5.25}$$

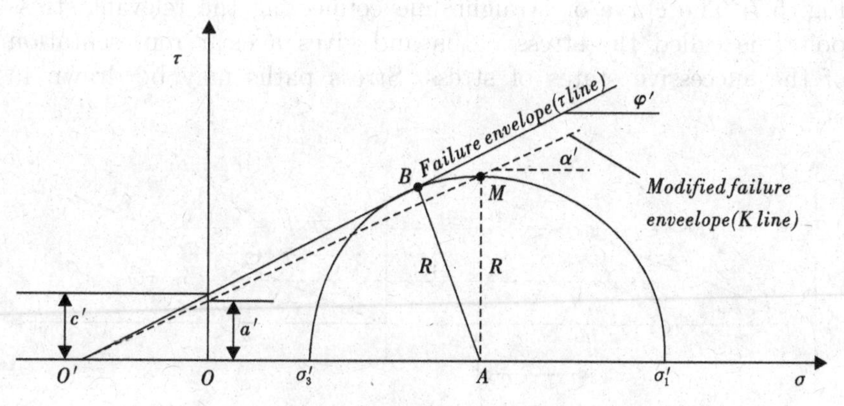


Fig. 5.6. Relation between $\tau$ line and $K$ line.

## 5.3 Shear Strength Tests

Various shear strength test procedures are possible but the three main types of test are as follows:

- **Unconsolidated–Undrained (UU):** The specimen is subjected to a specified all-round pressure and then the principal stress difference is applied immediately, with no drainage being permitted at any stage of the test.
- **Consolidated–Undrained (CU):** Drainage of the specimen is permitted under a specified all-round pressure until the consolidation is complete; the principal stress difference is then applied with no drainage being permitted. Pore water pressure measurements may be made during the undrained part of the test.
- **Consolidated–Undrained (CD):** Drainage of the specimen is permitted under a specified all-round pressure until the consolidation is complete; with drainage still being permitted, the principal stress difference is then applied at a rate slow enough to ensure that the excess pore water pressure is maintained at zero.

## 5.4 Effective Stress Paths

The successive states of stress in a test specimen can be represented by a series of Mohr circles or by a series of stress points as shown in Fig. 5.7. The curve or straight line connecting the relevant stress points is called the stress paths and gives a clear representation of the successive states of stress. Stress paths may be drawn in

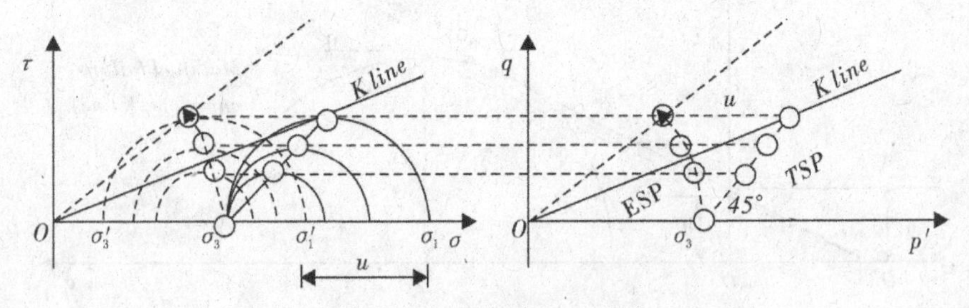


Fig. 5.7. Relationship of Mohr circle and stress path

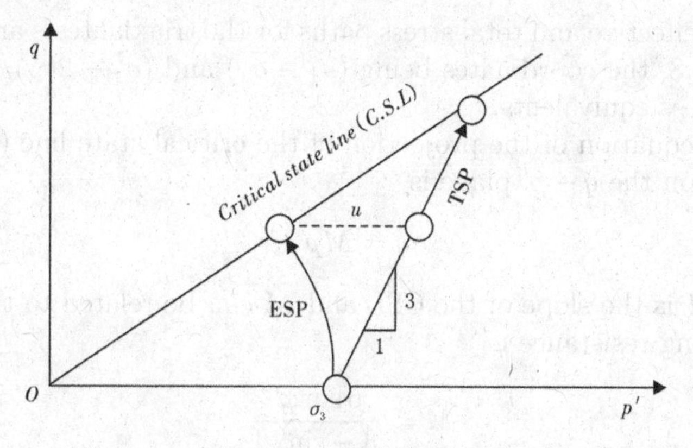


Fig. 5.8. Stress path.

terms of either the effective or total stress. The horizontal distance between the effective and total stress paths is the value of pore water pressure. The effective and total stress paths (denoted by ESP and TSP, respectively) for the triaxial tests are shown in Fig. 5.8, the coordinates being $(\sigma_1 - \sigma_3)/2$ and $(\sigma_1' + \sigma_3')/2$ or the total stress equivalents.

$$q = \frac{\sigma_1' - \sigma_3'}{2}, \tag{5.26}$$

$$p' = \frac{\sigma_1' + \sigma_3'}{2}. \tag{5.27}$$

The effective stress paths terminate on the modified failure envelope. The effective stress paths for the drained tests and all the total stress paths are straight lines at a slope of 45°. The detailed shape of the effective stress paths for the CU tests depends on the pore water pressure $u$.

Stress paths are also plotted with respect to deviator stress $(\sigma_1 - \sigma_3)$ and average effective principal stress $(\sigma_1' + \sigma_2' + \sigma_3')/3$, denoted by $q$ and $p'$, respectively. In the triaxial test, the intermediate principal stress $(\sigma_2')$ is equal to the minor principal stress $(\sigma_3')$; therefore,

$$q = \sigma_1' - \sigma_3', \tag{5.28}$$

$$p' = \frac{\sigma_1' + 2\sigma_3'}{3} = \sigma_3 + \frac{1}{3}q - u. \tag{5.29}$$

The effective and total stress paths for the triaxial tests are shown in Fig. 5.8, the coordinates being $(\sigma_1 - \sigma_3)$ and $(\sigma_1' + 2\sigma_3')/3$ or the total stress equivalents.

The equation of the projection of the critical state line (CSL) in Fig. 5.8 on the $q - p'$ plane is

$$q = Mp', \tag{5.30}$$

where $M$ is the slope of the CSL and $\mathbf{M}$ can be related to the angle of shearing resistance $\varphi'$

$$M = \frac{6 \sin \varphi'}{3 - \sin \varphi'}. \tag{5.31}$$

## 5.5 Characteristics of Shear Strength of Cohesionless Soil

The main factor influencing the shear strength of sand is the initial compactness. The initial compactness could be reflected by the size of void ratio. In general, the smaller the initial void ratio and the closer contact of particles, the greater sliding and rolling friction between the particles and the greater friction resistance. Thus, the shear strength is greater.

Figure 5.9 shows the stress–strain relationship and volume change of the sand under the same confining pressure $\sigma_3$ with the different initial void ratio during shearing. Obviously, the void ratio of compacted dense sand is smaller, and there is an obvious peak on stress–strain relationship curve. The stress begins to decrease with the increase of the strain after the peak. This feature is called the strain softening. The strength of the peak is called the peak strength, and the strength of the final stable value is called the residual strength. When the dense sand shears, its volume decreases slightly in the beginning and then increases significantly (dilatancy). Generally, the volume is greater than the initial volume. This feature is called the dilatancy of soil. The strength of the loose sand increases with the increase of axial strain, and generally, there is no peak on the curve. This feature is called the strain hardening. When the loose sand shears, the particles scroll to the equilibrium position and arrange more closely, following which the volume decreases. This feature is called the shear shrinkage. With the increase of pressure

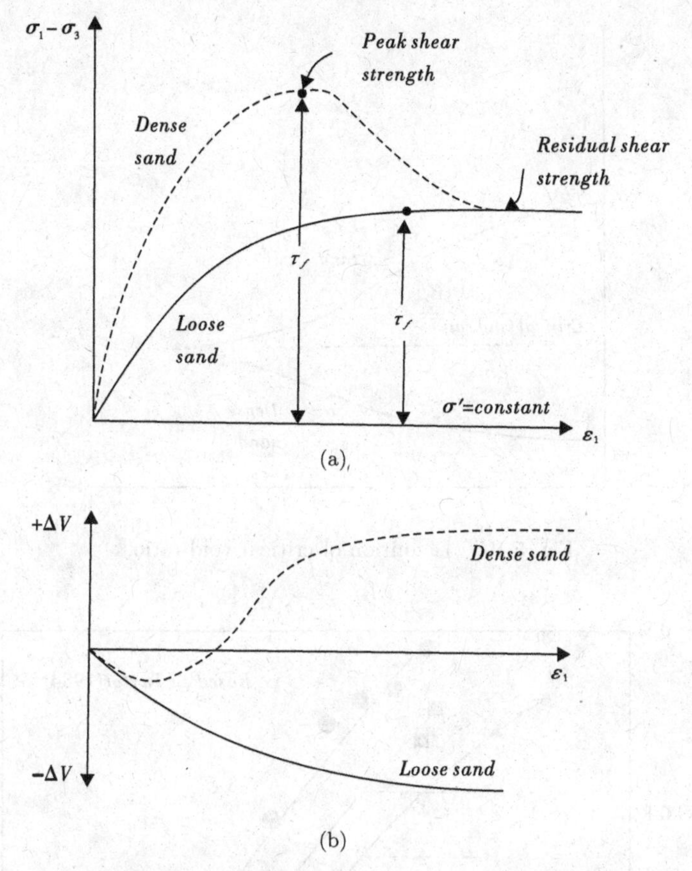


Fig. 5.9. Experimental results of sand. (a) Relationship of stress and strain and (b) relationship of volume and strain.

and the crush of soil particles, the dilatancy trend of dense sand will gradually disappear. So under the high confining pressure, no matter how elastic the sand is, shear shrinkage will occur.

For the shear tests of dense sands, there is a tendency of the specimen to dilate as the test progresses. Similarly, in loose sand the volume gradually decreases. An increase or decrease of volume means a change in the void ratio of soil. The nature of the change of the void ratio with strain for loose and dense sands in shown in Fig. 5.10. The void ratio for which the change of volume remains constant during shearing is called the critical void ratio. Figure 5.11 shows the results of a few drained triaxial tests on washed Fort Peck sand.

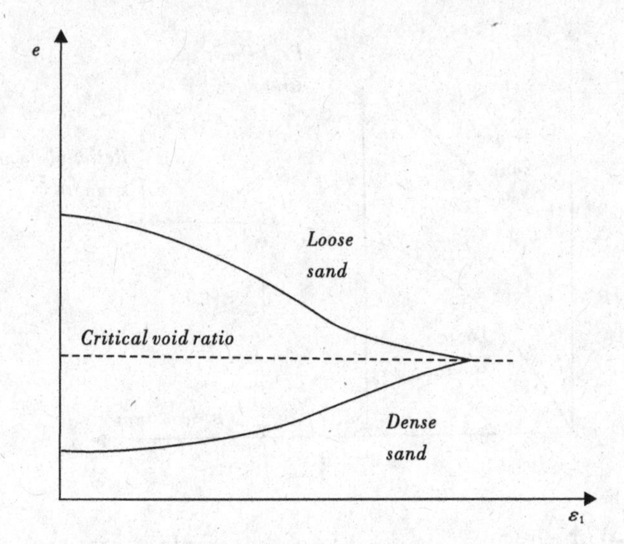


Fig. 5.10. Definition of critical void ratio.

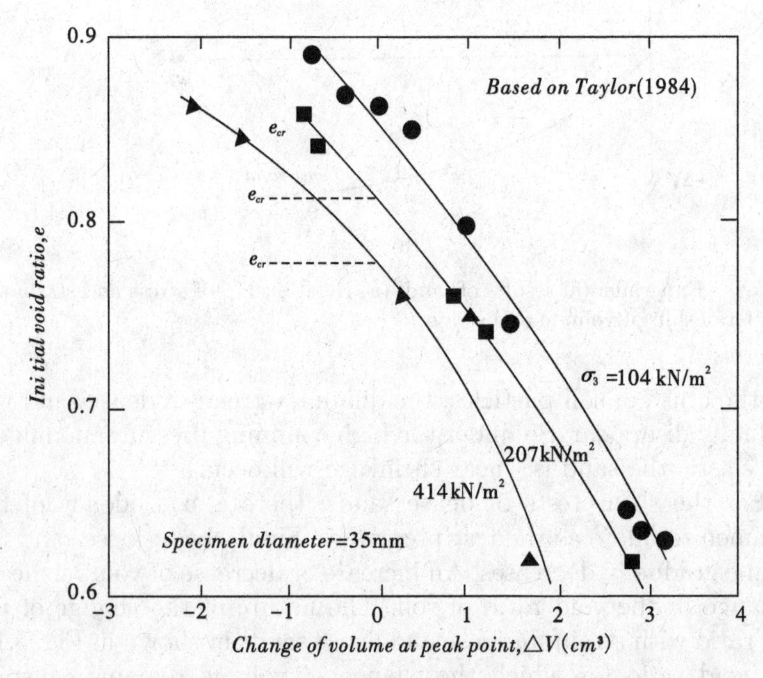


Fig. 5.11. Critical void ratio from triaxial test on Fort Peck sand.

The void ratio after the application of $\sigma_3$ is plotted in the ordinate, and the change of volume, $V$, at the peak point of the stress–strain plot, is plotted along the abscissa. For a given $\sigma_3$, the void ratio corresponding to $V = 0$ is the critical void ratio. Note that the critical void ratio is a function of the confining pressure $\sigma_3$. It is, however, necessary to recognize that, whether the volume of the soil specimen increases or decreases, the critical void ratio reaches only in the shearing zone, even if it is generally calculated on the basis of the total volume change of the specimen. The concept of the critical void ratio was first introduced in 1938 by Casagrande to study liquefaction of granular soils. When a natural deposit of saturated sand that has a void ratio greater than the critical void ratio is subjected to a sudden shearing stress (due to an earthquake or to blasting, for example), the sand will undergo a decrease in volume. This will result in an increase of pore water pressure $u$. At a given depth, the effective stress is given by the relation $\sigma' = \sigma - u$. If $\sigma$ (i.e. the total stress) remains constant and $u$ increases, the result will be a decrease in $\sigma'$. This, in turn, will reduce the shear strength of the soil. If the shear strength is reduced to a value which is less than the applied shear stress, the soil will fail. This is called soil liquefaction.

The shear strength of the sand soil is determined by the normal effective stress and the internal friction angle. The internal friction angle of the dense sand is relevant to the initial void ratio, roughness on the surface of the soil particles, grain composition, and so on. If the initial void ratio is small and the surface of soil particles is rough, then the internal friction angle of the well-graded soil is greater. The internal friction angle of the loose sand more or less equals the natural rest angle of dry sand, which can be measured in the lab by an easy way.

## 5.6 Characteristics of Shear Strength of Cohesive Soil

The shear strength of cohesive soils can generally be determined in the laboratory by either the direct shear test equipment or the triaxial shear test equipment. The triaxial test is more commonly used. Only the shear strength of saturated cohesive soils will be treated here. The shear strength based on the effective stress can be given by $\tau = \sigma' \tan \varphi' + c'$. For normally consolidated clays, $c \approx 0$, and for overconsolidated clays, $c > 0$.

Three conventional test types are conducted in the laboratory:

(1) Consolidated drained test (CD test).
(2) Consolidated undrained test (CU test).
(3) Unconsolidated undrained test (UU test).

Each of these tests will be separately considered in the following sections.

### 5.6.1 *Consolidated drained test*

For the consolidated drained test, the saturated soil specimen is first subjected to a confining pressure through the chamber fluid. The connection to the drainage is kept open for complete drainage, so that the pore water pressure is equal to zero. Then the deviator stress increases at a very slow rate, keeping the drainage valve open to allow complete dissipation of the resulting pore water pressure. Figure 5.12 shows the nature of the variation of the deviator stress with axial strain. From Fig. 5.12, it must also be pointed out that, during the application of the deviator stress, the volume of the specimen gradually reduces for normally consolidated clays. However, the overconsolidated clays go through some reduction of volume initially and then expand. In a consolidated drained test, the total stress is equal to the effective stress, since the excess pore water pressure is zero.

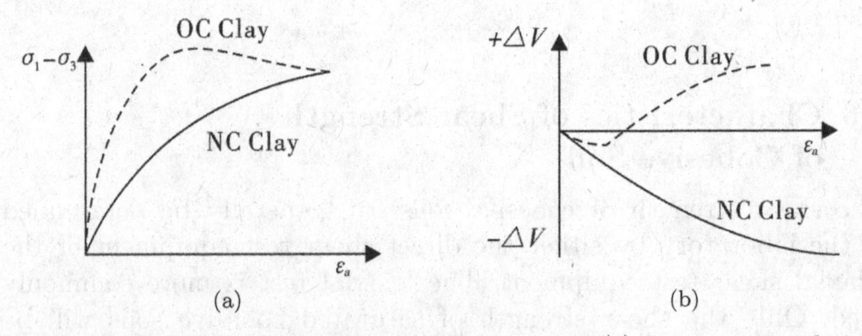


Fig. 5.12. Consolidated drained triaxial test in clay. (a) Stress–strain relationship and (b) volumetric and axial strain relationship.

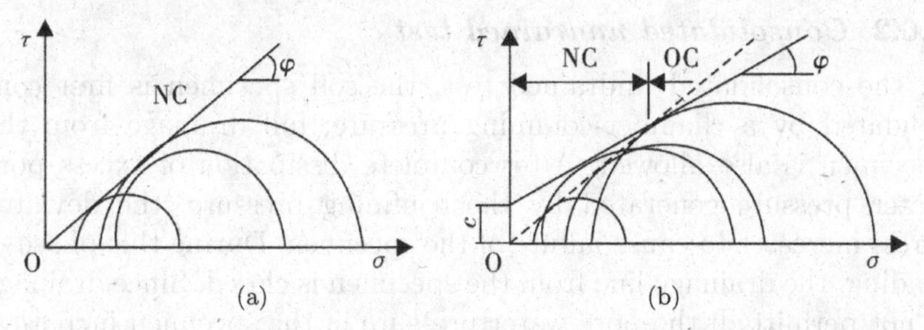


Fig. 5.13. Failure envelope for drained test. (a) Normally consolidated clay and (b) overconsolidated clay.

From the results of a number of tests conducted using several specimens, Mohr's circles at failure can be plotted as shown in Fig. 5.13. The values of $c$ and $\varphi$ are obtained by drawing a common tangent to Mohr's circles, which is the Mohr–Coulomb envelope. For the normally consolidated clays, $c = 0$. Thus, the equation of the Mohr–Coulomb envelope can be given by $\tau_f = \sigma' \tan\varphi$. The slope of the failure envelope will give us the angle of friction of the soil as follows:

$$\sin\varphi = \left(\frac{\sigma_1' - \sigma_3'}{\sigma_1' + \sigma_3'}\right)_{\text{failure}}. \tag{5.32}$$

For the overconsolidated clays (Fig. 5.13), $c \neq 0$. So the shear strength follows the equation $\tau_f = \sigma' \tan\varphi + c$. The values of $c$ and $\varphi$ can be determined by measuring the intercept of the failure envelope on the shear stress axis and the slope of the failure envelope, respectively. The following formula is thus obtained:

$$\sin\varphi = \frac{\sigma_1' - \sigma_3'}{(\sigma_1' + \sigma_3') + c\cot\varphi}. \tag{5.33}$$

Figure 5.12 shows that the deviator stress reaches a constant value at very large strains. The shear strength of clays at very large strains is referred to as the residual shear strength (i.e. the ultimate shear strength). It has been proved that the residual strength of a given soil is independent of past stress history. The residual friction angle in clays is importance in subjects such as the long-term stability of slopes.

### 5.6.2 *Consolidated undrained test*

In the consolidated undrained test, the soil specimen is first consolidated by a chamber-confining pressure; full drainage from the specimen is also allowed. After complete dissipation of excess pore water pressure generated by the confining pressure, the deviator stress increases to cause failure of the specimen. During this phase of loading, the drainage line from the specimen is closed. Since drainage is not permitted, the pore water pressure in the specimen increases. Figure 5.14 shows the results of the consolidated undrained test. The pore water pressure at failure $u_f$ is positive for normally consolidated clays and becomes negative for overconsolidated clays. Thus, $u_f$ is dependent on the overconsolidation ratio. The overconsolidation ratio, OCR, for triaxial test conditions may be defined as

$$OCR = \frac{\sigma_3'}{\sigma_3}, \tag{5.34}$$

where $\sigma_3' = \sigma_3$ is the maximum chamber pressure at which the specimen is consolidated and then allowed to rebound under a chamber pressure of $\sigma_3$.

Consolidated undrained tests on a number of specimens can be conducted to determine the shear strength parameters of a soil, as shown for the case of normally consolidated clay in Fig. 5.15. The Mohr's circle of the total-stress is shown by the solid line. The Mohr's circle of the effective stress is shown by a dashed line. A common tangent drawn to the effective-stress circle will give the Mohr–Coulomb failure envelope given by the equation $\tau = \sigma' \tan \varphi$.

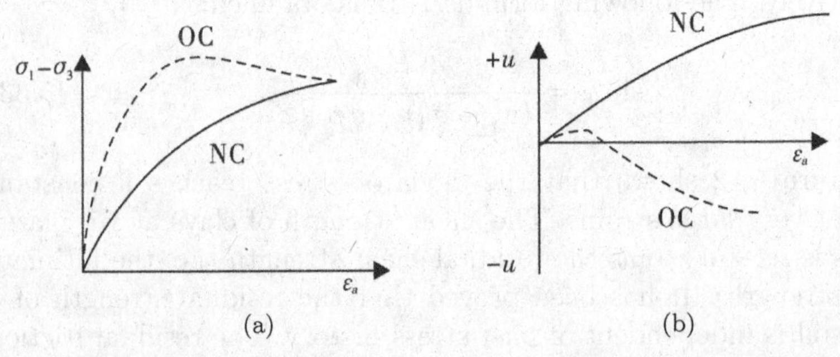


Fig. 5.14. Results of consolidated undrained test. (a) Stress–strain relationship and (b) pore pressure–strain relationship.

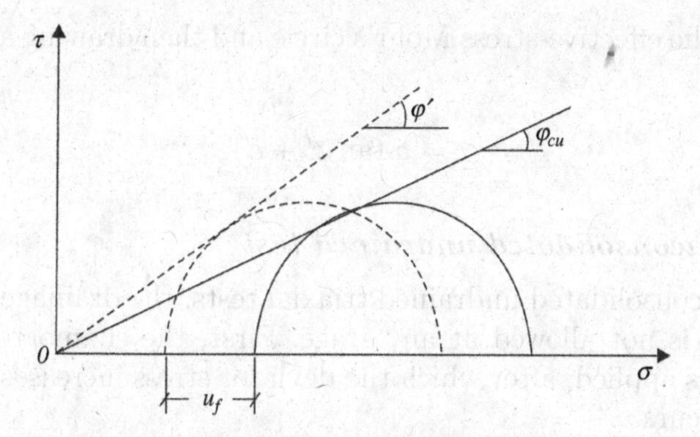


Fig. 5.15. Failure envelope for normally consolidated clay.

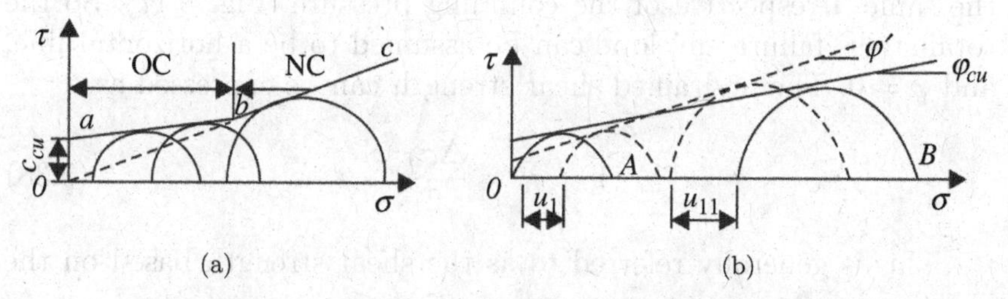


Fig. 5.16. Failure envelope for overconsolidated clay.

If we draw a common tangent to the total-stress circles, it will be a straight line passing through the origin. This is the total-stress failure envelope, and it may be given by

$$\tau = \sigma \tan \varphi_{cu}, \tag{5.35}$$

where $\varphi_{cu}$ is the consolidated undrained angle of friction.

The total stress failure envelope for an overconsolidated clay will be of the nature shown in Fig. 5.16 and can be given by the relation

$$\tau = \sigma \tan \varphi_{cu} + c_{cu}, \tag{5.36}$$

where $c_{cu}$ is the intercept of the total-stress failure envelope along the shear stress axis.

The shear strength parameters for the overconsolidated clay based on the effective stress, i.e. $c'$ and $\varphi'$, can be obtained by plotting the effective-stress Mohr's circle and then drawing a common tangent.

$$\tau = \sigma \tan \varphi' + c'. \tag{5.37}$$

### 5.6.3 *Unconsolidated undrained test*

In the unconsolidated undrained triaxial tests, the drainage from the specimen is not allowed at any stage. First, the chamber-confining pressure is applied, after which the deviator stress increases until the failure occurs.

The test type can be performed quickly, since drainage is not allowed. For a saturated soil, the deviator stress failure is practically the same, irrespective of the confining pressure (Fig. 5.17). So the total-stress failure envelope can be assumed to be a horizontal line, and $\varphi = 0$. The undrained shear strength can be expressed as

$$\tau = c_u = \frac{\Delta \sigma_f}{2}. \tag{5.38}$$

This is generally referred to as the shear strength based on the $\varphi = 0$ concept.

The fact that the strength of saturated clays in unconsolidated, undrained loading conditions is the same, irrespective of the confining pressure, can be explained as follows. For the saturated soil, $B = 1$ under undrained conditions, the pore water increases $\Delta\sigma_f$ as subjected to an additional confining pressure $\Delta\sigma_f$, while the effective stress is kept a constant. For a series of tests, there is only one effective stress circle, so only the undrained shear strength is measured.

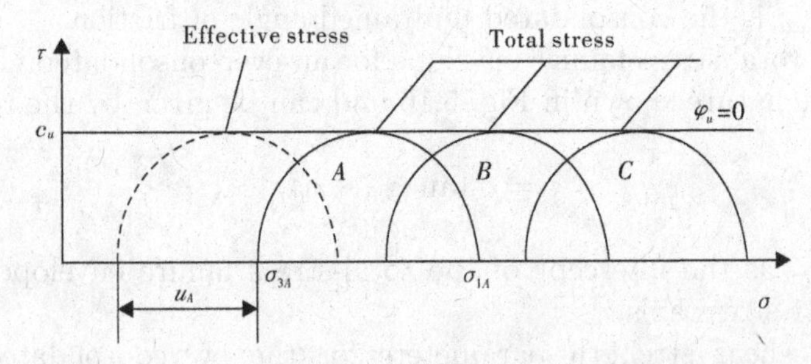


Fig. 5.17. Failure envelope of unconsolidated undrained triaxial tests.

According to the above analysis, it can be found that, for the total stresses, the strength difference due to the difference of the experiment method can be reflected by strength parameters, i.e. the total stress strength parameters contain the effect of the pore water pressure. The total friction angle satisfies the following equation: $\varphi_d > \varphi_{cu} > \varphi_u$. For the effective stresses, the effective strength parameters are almost the same.

## Exercises

5.1 The results of the triaxial compression test of various soil samples are as follows:

| $\sigma$ (kPa) | 50 | 200 | 200 | 300 |
|---|---|---|---|---|
| $\tau$(kPa) | 23.4 | 36.7 | 63.9 | 90.8 |

(1) Calculate the internal friction angle and cohesion of the soil.
(2) When $\sigma = 280\,\text{kPa}$ and $\tau = 80\,\text{kPa}$, determine the failure state of the soil.

5.2 A direct shear test is conducted on sand. The failure occurs at $\sigma = 100\,\text{kPa}$ and $\tau = 60\,\text{kPa}$.

(1) Calculate the internal friction angle.
(2) If $\sigma = 250\,\text{kPa}$, calculate the shear strength of sand.
(3) Calculate $\sigma_1$ and $\sigma_3$.

5.3 The strength parameters of soil are $c = 20\,\text{kPa}$ and $\varphi = 22°$, and the normal stress and shear stress acting on a slope plane are $\sigma = 100\,\text{kPa}$ and $\tau = 60.4\,\text{kPa}$, respectively. Determine the failure or stability of soil along the plane.

5.4 A series of conventional triaxial consolidated drained tests are conducted on sands. The principal stress difference is $\sigma_1 - \sigma_3 = 400\,\text{kPa}$, and the confining pressure is $\sigma_3 = 100\,\text{kPa}$. Determine the strength parameters of sand.

## Bibliography

R. F. Craig (2004). *Craig's Soil Mechanics* (Seventh edition). CRC Press.
B. M. Das (2008). *Advanced Soil Mechanics* (Third edition). Taylor and Francis.
H. Liao (2018). *Soil Mechanics* (Third Edition). Higher Education Press, Beijing.
K. Terzaghi, R. B. Peck, and G. Mesri (1996). *Soil Mechanics in Engineering Practice* (Third Edition). John Wiley & Sons, Inc.

Chapter 6

# Bearing Capacity

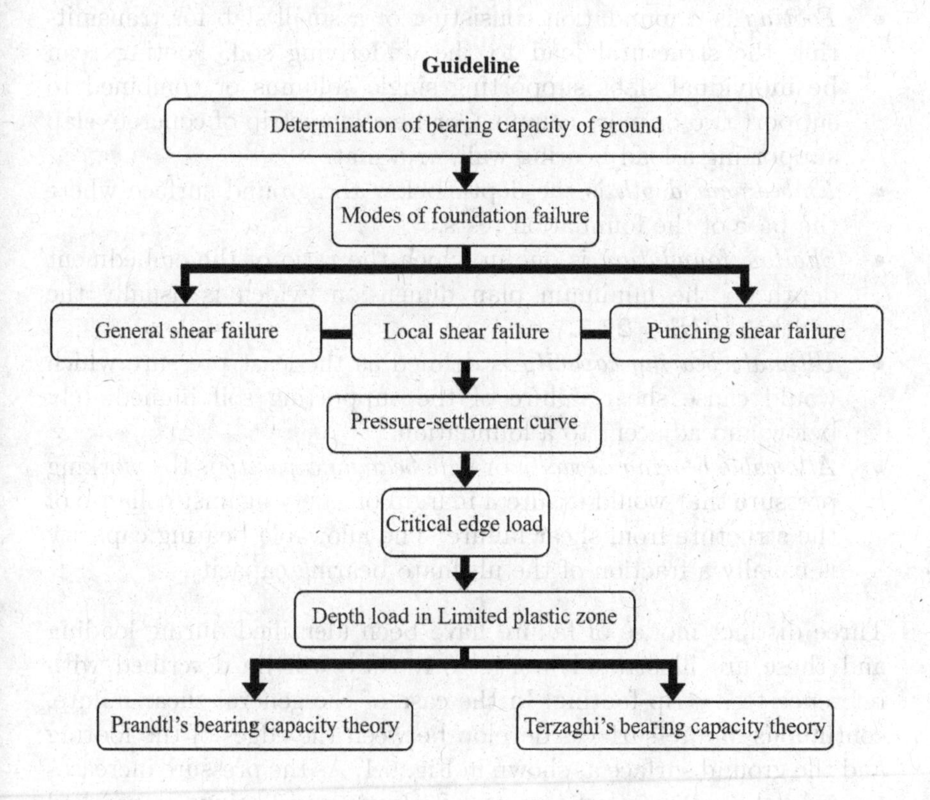

## 6.1 Introduction

In this chapter, bearing capacity of soils will be considered. Loads from a structure are transferred to the soil through a foundation. The limit equilibrium method of analysis is considered. The limit equilibrium method is used to find solutions for a variety of problems including bearing capacity of foundations, stability of retaining walls, and slopes.

Before considering these bearing capacity equations, the following definition key terms should be identified:

- *Foundation* is a structure that transmits loads to the underlying soils.
- *Footing* is a foundation consisting of a small slab for transmitting the structural load to the underlying soil. Footings can be individual slabs supporting single columns or combined to support two or more columns, or be a long strip of concrete slab supporting a load bearing wall, or a mat.
- *Embedment depth* is the depth below the ground surface where the base of the foundation rests.
- *Shallow foundation* is one in which the ratio of the embedment depth to the minimum plan dimension, which is usually the width, is $d/B = 2:5$.
- *Ultimate bearing capacity* is defined as the least pressure which would cause shear failure of the supporting soil immediately below and adjacent to a foundation.
- *Allowable bearing capacity* or *safe bearing capacity* is the working pressure that would ensure a margin of safety against collapse of the structure from shear failure. The allowable bearing capacity is usually a fraction of the ultimate bearing capacity.

Three distinct modes of failure have been identified during loading and these are illustrated in Fig. 6.1: they will be described with reference to a strip footing. In the case of the general shear failure, continuous failure surfaces develop between the edges of the footing and the ground surface as shown in Fig. 6.1. As the pressure increases towards the value $q_f$ the state of plastic equilibrium is reached initially in the soil around the edges of the footing, and then gradually spreads downwards and outwards. Ultimately the state of plastic equilibrium is fully developed throughout the soil above the failure

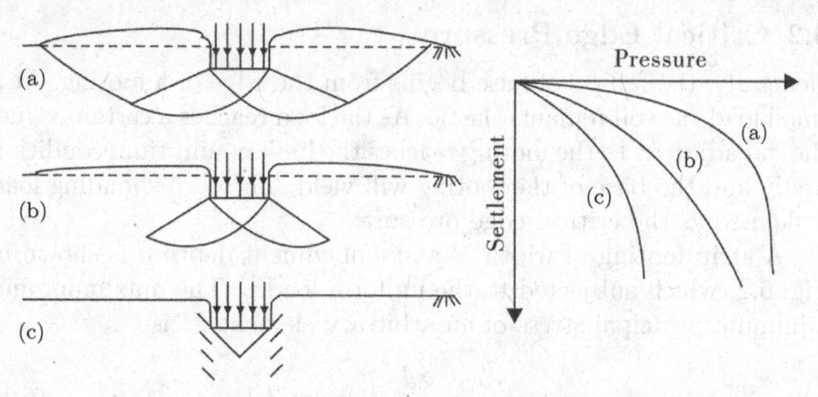


Fig. 6.1. Modes of failures: (a) general shear (b) local shear, and (c) punching shear.

surfaces. Heaving of the ground surface occurs on both sides of the footing although the final slip movement would occur only on one side, accompanied by tilting of the footing. This mode of failure is typical of soils of low compressibility (i.e. dense or stiff soils) and the pressure–settlement curve is of the general form shown in Fig. 6.1, the ultimate bearing capacity being well defined.

In the mode of local shear failure, there is significant compression of the soil under the footing and only partial development of the state of plastic equilibrium. The failure surfaces, therefore, do not reach the ground surface and only slight heaving occurs. Tilting of the foundation would not be expected. Local shear failure is associated with the high compressibility of soils, as indicated in Fig. 6.1, is characterized by the occurrence of relatively large settlements (which would be unacceptable in practice). The ultimate bearing capacity is not clearly defined.

The punching shear failure occurs when there is compression of the soil under the footing, accompanied by shearing in the vertical direction around the edges of the footing. There is no heaving of the ground surface away from the edges and no tilting of the footing. Relatively large settlements are also a characteristic of this mode and the ultimate bearing capacity is not well defined. The punching shear failure will also occur in a soil with low compressibility if the foundation is located at a considerable depth. In general the mode of failure depends on the compressibility of the soil and the depth of the foundation relative to its breadth.

## 6.2 Critical Edge Pressure

Generally, the failure of base begins from the edge of a footing. At a small load the soil remains elastic. As the load reaches a certain value, the soil adjacent to the footing reaches the limit equilibrium condition firstly and the base of the footing will yield. The corresponding load is defined as the critical edge pressure.

A strip footing of width $B$ and embedment depth $d$ is shown in Fig. 6.2, which subjected to the uniform load $p$. The maximum and minimum principal stress of an arbitrary element $M$ is

$$\sigma_{1,3} = \frac{p - \gamma_0 d}{\pi}(\beta_0 \pm \sin\beta_0), \tag{6.1}$$

where $\gamma_0$ is the average weight above the base (kN/m$^3$); $\beta_0$ is the angle of $M$ to two endpoints of uniform load (°).

The gross stress of $M$ is the sum of the gravity stress and surcharge stress. For simplicity, it is assumed that the gravity stress field is regards as the hydrostatic stress field, i.e., the lateral pressure coefficient equals to 1.0. So the total stress of $M$ is given by

$$\sigma_{1,3} = \frac{p - \gamma_0 d}{\pi}(\beta_0 \pm \sin\beta_0) + \gamma_0 d + \gamma z, \tag{6.2}$$

where $\gamma$ is the average weight below the base (kN/m$^3$).

When the element $M$ reaches the limit equilibrium condition, according to the Mohr–Coulomb criterion, the following principal

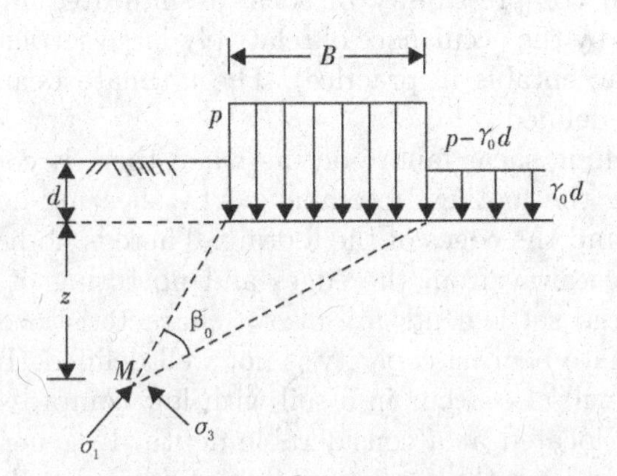


Fig. 6.2. The principal stresses at uniform load.

stresses exist:

$$\frac{\sigma_1 - \sigma_3}{\sigma_1 + \sigma_3 + 2c \cdot \cot\varphi} = \sin\varphi. \tag{6.3}$$

Substituting (6.2) into the above equation, the depth of plastic zone is obtained

$$z = \frac{p - \gamma_0 d}{\gamma\pi}\left(\frac{\sin\beta_0}{\sin\varphi} - \beta_0\right) - \frac{c}{\gamma}\cot\varphi - d\frac{\gamma_0}{\gamma}, \tag{6.4}$$

where $c$ is the cohesion of soil below base (kPa), $\varphi$ is the friction angle of soil below base (°).

Equation (6.4) is the boundary equation of plastic zone, which reflects the relationship between $z$ and $\beta_0$ of boundary of plastic zone. If $d$, $p$, $\varphi$, $\gamma$ and $c$ are known, the plastic zone boundary can be drawn.

The maximum depth of plastic zone can be obtained by differentiating Eq. (6.4)

$$\beta_0 = \frac{\pi}{2} - \varphi. \tag{6.5}$$

Substituting Eq. (6.5) into Eq. (6.4), the maximum depth is derived

$$z_{\max} = \frac{p - \gamma_0 d}{\gamma\pi}\left(\cot\varphi + \varphi - \frac{\pi}{2}\right) - \frac{c}{\gamma}\cot\varphi - d\frac{\gamma_0}{\gamma}. \tag{6.6}$$

According to the above equation, the critical edge pressure is

$$p_{cr} = \gamma_0 d N_q + c N_c, \tag{6.7}$$

where

$$N_q = 1 + \frac{\pi}{\cot + \varphi - \frac{\pi}{2}}, \tag{6.8a}$$

$$N_c = \frac{\pi\cot\varphi}{\cot\varphi + \varphi - \frac{\pi}{2}}. \tag{6.8b}$$

It can be found that $p_{cr}$ is dependent on $\varphi$, $\gamma_0$, $c$ and $d$, and independent of the width of footing.

A large number of engineering practices have shown that $p_{cr}$, as the bearing capacity design value, is conservative. The experiences have shown that in most cases, even if there is a certain range of plastic zones which do not exceed a permissible range, the structures can be kept in normal use. It is generally believed that the maximum depth of the plastic zone can reach the 1/4 of the width of the footing under a centric load. Under a small eccentric load, the maximum depth of the plastic zone is allowed to reach 1/3 of the width of footing. According to the conditions, the corresponding pressures ($p_{1/4}$ and $p_{1/3}$) can be obtained by substituting $z_{\max} = 1/4B$ and $z_{\max} = 1/3B$ into Eq. (6.6), respectively,

$$p_{1/4} = \frac{1}{2}\gamma B N_{1/4} + \gamma_0 d N_q + c N_c, \tag{6.9a}$$

$$p_{1/3} = \frac{1}{2}\gamma B N_{1/3} + \gamma_0 d N_q + c N_c, \tag{6.9b}$$

Table 6.1. The value of $N_{1/4}$, $N_{1/3}$, $N_q$, $N_c$ with $\varphi$.

| $\varphi$ | $N_{1/4}$ | $N_{1/3}$ | $N_q$ | $N_c$ |
|---|---|---|---|---|
| 0 | 0 | 0 | 1.0 | 3.14 |
| 2 | 0.06 | 0.08 | 1.12 | 3.32 |
| 4 | 0.12 | 0.16 | 1.25 | 3.51 |
| 6 | 0.20 | 0.27 | 1.40 | 3.71 |
| 8 | 0.28 | 0.37 | 1.55 | 3.93 |
| 10 | 0.36 | 0.48 | 1.73 | 4.17 |
| 12 | 0.46 | 0.60 | 1.94 | 4.42 |
| 14 | 0.60 | 0.80 | 2.17 | 4.70 |
| 16 | 0.72 | 0.96 | 2.43 | 5.00 |
| 18 | 0.86 | 1.15 | 2.72 | 5.31 |
| 20 | 1.00 | 1.33 | 3.10 | 5.66 |
| 22 | 1.20 | 1.60 | 3.44 | 6.04 |
| 24 | 1.40 | 1.86 | 3.87 | 6.45 |
| 26 | 1.60 | 2.13 | 4.37 | 6.90 |
| 28 | 2.00 | 2.66 | 4.93 | 7.40 |
| 30 | 2.40 | 3.20 | 5.60 | 7.95 |
| 32 | 2.80 | 3.73 | 6.35 | 8.55 |
| 34 | 3.20 | 4.26 | 7.20 | 9.22 |
| 36 | 3.60 | 4.80 | 8.25 | 9.97 |
| 38 | 4.20 | 5.60 | 9.44 | 10.80 |
| 40 | 5.00 | 6.66 | 10.84 | 11.73 |
| 42 | 5.80 | 7.73 | 12.70 | 12.80 |
| 44 | 6.40 | 8.52 | 14.50 | 14.00 |
| 45 | 7.40 | 9.86 | 15.60 | 14.60 |

where

$$N_{1/3} = \frac{\pi}{3\left(ctg\varphi + \varphi - \frac{\pi}{2}\right)}, \quad (6.10a)$$

$$N_{1/4} = \frac{\pi}{4\left(ctg\varphi + \varphi - \frac{\pi}{2}\right)}, \quad (6.10b)$$

where $N_{1/4}$, $N_{1/3}$, $N_q$, $N_c$ are the coefficients of bearing capacity, which can be determined according to Table 6.1.

## 6.3 Prandtl's Bearing Capacity Theory

Prandtl (1920) showed theoretically that a wedge of material is trapped below a rigid strip footing when it is subjected to a concentric load. A suitable failure mechanism for a strip footing is shown in Fig. 6.3. The footing, with width $B$ and infinite length, carries a uniform pressure $p$ on the surface of a mass of homogeneous, isotropic soil. The shear strength parameters for the soil are $c$ and $\varphi$ but the unit weight is assumed to be zero.

When the pressure becomes equal to the ultimate bearing capacity $q_f$, the state of plastic equilibrium, in the form of an active Rankine zone, below the footing, the angles ABC and BAC being $(45° + \frac{\varphi}{2})$. The downward movement of the wedge ABC forces the adjoining soil sideways, producing outward lateral forces on both sides of the wedge. The passive Rankine zones ADF and BGE therefore develop on both sides of the wedge ABC, the angles DFA and EGB being $(45° - \frac{\varphi}{2})$. The transition between the downward movement of the wedge ABC and the lateral movement of the wedges ADF and BGE take place through zones of radial shear (also known as slip fans) CDF and CEG, the surfaces EC and CD being

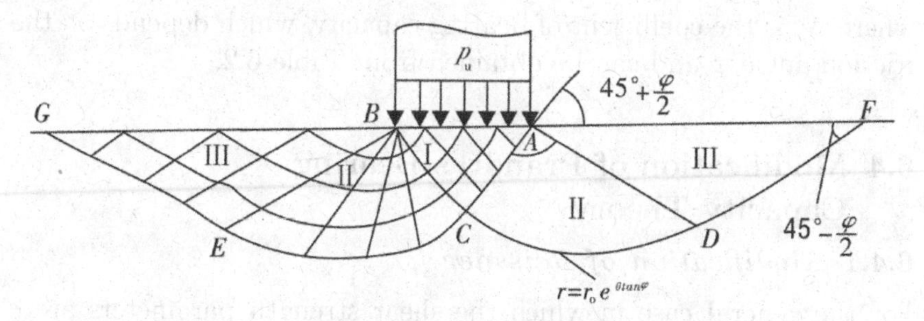


Fig. 6.3. General failure under strip footing (Prandtl).

Table 6.2. The coefficient of bearing capacity.

| $\varphi^0$ | $N_\gamma$ | $N_q$ | $N_c$ | $\varphi^0$ | $N_\gamma$ | $N_q$ | $N_c$ |
|---|---|---|---|---|---|---|---|
| 0 | 0 | 1.00 | 5.14 | 24 | 6.90 | 9.61 | 19.3 |
| 2 | 0.01 | 1.20 | 5.69 | 26 | 9.53 | 11.9 | 22.3 |
| 4 | 0.05 | 1.43 | 6.17 | 28 | 13.1 | 14.7 | 25.8 |
| 6 | 0.14 | 1.72 | 6.82 | 30 | 18.1 | 18.4 | 30.2 |
| 8 | 0.27 | 2.06 | 7.52 | 32 | 25.0 | 23.2 | 35.5 |
| 10 | 0.47 | 2.47 | 8.35 | 34 | 34.5 | 29.5 | 42.2 |
| 12 | 0.76 | 2.97 | 9.29 | 36 | 48.1 | 37.8 | 50.6 |
| 14 | 1.16 | 3.58 | 10.4 | 38 | 67.4 | 48.9 | 61.4 |
| 16 | 1.72 | 4.33 | 11.6 | 40 | 95.5 | 64.2 | 75.4 |
| 18 | 2.49 | 5.25 | 13.1 | 42 | 137 | 85.4 | 93.7 |
| 20 | 3.54 | 6.40 | 14.8 | 44 | 199 | 115 | 118 |
| 22 | 4.96 | 7.82 | 16.9 | 45 | 241 | 134 | 133 |

logarithmic spirals (or circular arcs if $\varphi = 0$) to which AC and GE, or BC and FD, are tangential. A state of plastic equilibrium thus exists above the surface GECDF, the remainder of the soil mass being in a state of elastic equilibrium.

The following exact solution can be obtained, using the plasticity theory, for the ultimate bearing capacity of a strip footing in the surface of a weightless soil:

$$p_u = c\left[e^{\pi\tan\varphi}\cdot\tan^2\left(45^\circ+\frac{\varphi}{2}\right)-1\right]\cdot\cot\varphi = c\cdot N_c, \qquad (6.11)$$

where

$$N_c = \left[e^{\pi\tan\varphi}\cdot\tan^2\left(45^\circ+\frac{\varphi}{2}\right)-1\right]\cdot\cot\varphi, \qquad (6.12)$$

where $N_c$ is the coefficient of bearing capacity, which depends on the friction angle $\varphi$ and can be obtained from Table 6.2.

## 6.4 Modification of Prandtl's Bearing Capacity Theory

### 6.4.1 *Modification of Reissner*

For the general case in which the shear strength parameters are $c$ and $\varphi$ it is necessary to consider a surcharge pressure $q$ acting on the

soil surface would be zero. For this case, Reissner (1924) presented the solution due to the surcharge pressure $q$:

$$p_u = qe^{\pi \tan\varphi} \cdot \tan^2\left(45^\circ + \frac{\varphi}{2}\right) = qN_q, \tag{6.13}$$

where

$$N_q = e^{\pi \tan\varphi} \tan^2\left(45^\circ + \frac{\varphi}{2}\right), \tag{6.14}$$

where $N_q$ is the coefficient of bearing capacity, and can be obtained from Table 6.2.

Combining Eqs. (6.11) and (6.13), the ultimate bearing capacity is

$$p_u = \gamma_0 d N_q + cN_c. \tag{6.15}$$

### 6.4.2 *Modification of Taylor*

However, an additional term must be added to Eq. (6.15) to take the component of bearing capacity due to the self-weight of soil into account. Taylor (1940) proposed that if the gravity of soil is taken into account and the failure surface is the same as Prandtl's assumption, the shear strength increases accordingly. Taylor assumed that the incremental shear strength can be denoted by an equivalent cohesive force $c' = \gamma \cdot t \cdot \tan\varphi$ in which $t$ is the height of the sliding failure soil. So replacing $c$ with $c + c'$ in Eq. (6.15), the ultimate bearing capacity considering the self-weight of soil is

$$\begin{aligned} p_u &= qN_q + (c + c')N_c = qN_q + c'N_c + cN_c \\ &= qN_q + cN_c + \gamma \cdot \frac{B}{2} \tan\left(45^\circ + \frac{\varphi}{2}\right)\left[e^{\pi \tan\varphi} \tan^2\left(45^\circ + \frac{\varphi}{2}\right) - 1\right] \\ &= \frac{1}{2}\gamma \cdot BN_\gamma + qN_q + cN_c, \end{aligned} \tag{6.16}$$

where

$$N_\gamma = \tan\left(45^\circ + \frac{\varphi}{2}\right)\left[e^{\pi \tan\varphi} \tan^2\left(45^\circ + \frac{\varphi}{2}\right) - 1\right]$$

is obtained from Table 6.2.

## 6.5 Terzaghi's Bearing Capacity Theory

Terzaghi (1943) derived a bearing capacity equations based on Prandtl (1920) failure mechanism and the limit equilibrium method for a footing at a depth $d$ below the ground level of a homogeneous soil. Foundations are not normally located on the surface of a soil mass, as assumed in the above solutions; but at a depth $d$ below the surface as shown in Fig. 6.4. Terzaghi assumed the following:

(1) The soil is a semi-infinite, homogeneous, isotropic, weightless, rigid-plastic material.
(2) The embedment depth is less than the width of the footing ($d < B$).
(3) The general shear failure occurs.
(4) The angle $\theta$ in the wedge (Fig. 6.4) is $\varphi$. Later, it was found (Vesic, 1973) that $\theta = 45° + \varphi/2$.
(5) The shear strength of the soil above the footing base is negligible.
(6) The soil above the footing base can be replaced by a surcharge stress ($= \gamma_0 d$).
(7) The base of the footing is rough.

The ultimate bearing capacity of the soil under a shallow strip footing can be expressed by the following general equation:

$$p_u = \frac{1}{2}\gamma B N_\gamma + cN_c + qN_q, \tag{6.17}$$

where $N_\gamma$, $N_c$ and $N_q$ are the bearing capacity factors depending only on the value of $\varphi$. The first term in Eq. (6.17) represents the contribution to the bearing capacity resulting from the self-weight of the soil, the second term is the contribution of the constant component of shear strength and the third term is the contribution

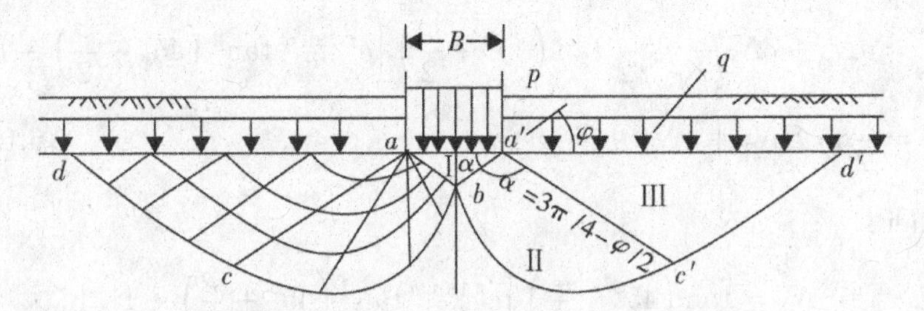


Fig. 6.4. Failure surface assumed by Terzaghi.

due to the surcharge pressure. It should be realized, however, that the superposition of the components of bearing capacity is theoretically incorrect for a plastic material. However, any resulting error is considered to be on the safe side.

For many years Terzaghi's bearing capacity factors were widely used. Terzaghi assumed that the angles $baa'$ and $a'ab$ in Fig. 6.4 were equal to $\varphi$ (i.e. $aba'$ was not considered to be an active Rankine zone). The Values of $N_\gamma$ were obtained by determining the total passive resistance and the adhesion force on the planes $ab$ and $a'b$. Terzaghi's values of $N_c$ and $N_q$ were obtained by modifying the Prandtl–Reissner solution. However, Terzaghi's values have now been largely superseded.

It is now considered that the values of $N_c$ and $N_q$ expressed in Eq. (6.17) should be used in bearing capacity calculations, i.e.

$$N_q = \frac{e^{\left(\frac{3}{2}\pi - \varphi\right)\tan\varphi}}{2\cos^2\left(45^0 + \frac{\varphi}{2}\right)}, \tag{6.18a}$$

$$N_c = (N_q - 1)\cot\varphi. \tag{6.18b}$$

For $N_\gamma$, Terzaghi did not given the solution, but plotted $N_\gamma$, $N_q$, $N_c$ with $\varphi$ as shown in Fig. 6.5. $N_\gamma$ can be determined according to the solid line.

Currently, various equations have been proposed for the $N_\gamma$ in the literature. Among the popular equations are

- Vesic (1973):

$$N_\gamma = 2(N_q + 1)\tan\varphi_p. \tag{6.19a}$$

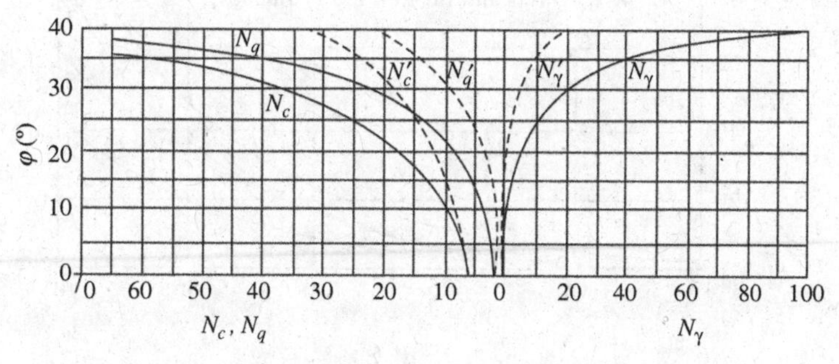


Fig. 6.5. The coefficient of Terzaghi's bearing capacity.

- Meyerhof (1976):

$$N_\gamma = (N_q - 1)\tan(1.4\varphi_p). \tag{6.19b}$$

- Davis and Booker (1971):

$$N_\gamma = 0.1054\exp(9.6\varphi_p) \quad \text{for rough footing,} \tag{6.19c}$$

$$N_\gamma = 0.0663\exp(9.3\varphi_p) \quad \text{for smooth footing.} \tag{6.19d}$$

The differences between these popular bearing capacity factors are shown in Fig. 6.6.

The bearing capacity factor, $N_\gamma$, proposed by Davis and Booker (1971) is based on a refined plasticity method and gives the conservative values compared with Vesic (1973). Meyerhof's $N_\gamma$ values are equal to Davis and Booker's $N_\gamma$ for $\varphi_p$ less than about 35°.

Equation (6.17) was derived under the condition of the general shear failure, which is suitable for the small compressibility soils. For the loose and large compressibility soil, the local shear failure may occur, in which the settlement is large and the ultimate bearing capacity is small. For the case, Eq. (6.17) can be modified by reducing the shear strength indexes $\varphi$, $c$.

$$c' = \frac{2}{3}c, \tag{6.20a}$$

$$\tan\varphi' = \frac{2}{3}\tan\varphi. \tag{6.20b}$$

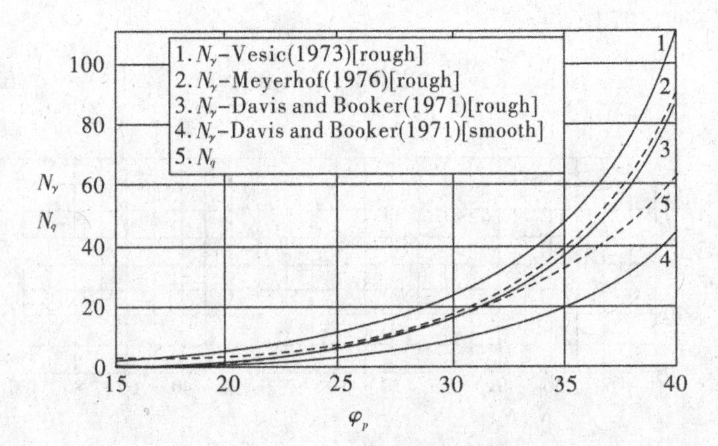


Fig. 6.6. Comparison of some bearing capacity factors.

Then the ultimate bearing capacity for local shear failure is

$$p_u = c'N'_c + qN'_q + \frac{1}{2}\gamma BN'_\gamma, \tag{6.21}$$

where $N'_q$, $N'_\gamma$ and $N'_c$ are obtained according to the dashed line in Fig. 6.5.

The problems involved in extending the two-dimensional solution for a strip footing to a three-dimensional one should be considered. Accordingly, the ultimate bearing capacities of square, rectangular and circular footings are determined by introducing the semi-empirical shape factors applied to the Solution for a strip footing. The bearing capacity factors $N_\gamma$, $N_c$ and $N_q$ should be multiplied by the respective shape factors $S_\gamma$, $S_c$ and $S_q$. The shape factors proposed by Terzaghi and Peck (1996) are still widely used in practice although they are considered to give conservative values of the ultimate bearing capacity for high values of $\varphi$. The factors are $S_\gamma = 0.8$ for a square footing or 0.6 for a circular footing, $S_c = 1.2$ and $S_q = 1$. Thus the ultimate bearing capacity becomes:

- For a square footing:

$$p_{us} = 0.4\gamma BN_\gamma + \gamma_0 dN_q + 1.2cN_c \quad \text{the general shear failure,} \tag{6.22}$$

$$p_{us} = 0.4\gamma BN'_\gamma + \gamma_0 dN'_q + 1.2c'N'_c \quad \text{the local shear.} \tag{6.23}$$

- For a circular footing:

$$p_{ur} = 0.3\gamma RN_\gamma + \gamma_0 dN_q + 1.2cN_c \quad \text{the general shear,} \tag{6.24}$$

$$p_{ur} = 0.3\gamma RN'_\gamma + \gamma_0 dN'_q + 1.2c'N'_c \quad \text{the local shear.} \tag{6.25}$$

where $R$ is the diameter of a circular footing.

It should be recognized that the results of the bearing capacity calculations are very sensitive to the values of the shear strength parameters, especially the higher values of $\varphi$. Therefore, the accuracy of the shear strength parameters must be considered.

Vesic (1973) considered the self-weight of soil and the embedment depth of footing based on Prandtl's theory, and gave the bearing capacity for strip footing:

$$p_u = cN_c + qN_q + \frac{1}{2}\gamma BN_\gamma, \tag{6.26}$$

where $N_c$, $N_q$, $N_\gamma$ are given by, respectively,

$$N_q = e^{\pi \tan\varphi} \tan^2\left(45^\circ + \frac{\varphi}{2}\right), \tag{6.27a}$$

$$N_c = (N_q - 1) \cdot \cot\varphi, \tag{6.27b}$$

$$N_\gamma = 2(N_q + 1)\tan\varphi. \tag{6.27c}$$

According to the factors influencing the bearing capacity, Vesic modified Eq. (6.26) and proposed many formulas of bearing capacity.

Equation (6.26) is only suitable for strip footing. For other shape footings, the following equation can be used:

$$p_u = cN_cS_c + qN_qS_q + \frac{1}{2}\gamma BN_\gamma S_\gamma, \tag{6.28}$$

where $S_c$, $S_q$, $S_\gamma$ are the coefficients of shape.

- For a rectangle footing:

$$\begin{cases} S_c = 1 + \dfrac{B}{l} \cdot \dfrac{N_q}{N_c} \\ S_q = 1 + \dfrac{B}{l} \cdot \tan\varphi \\ S_\gamma = 1 - 0.4\dfrac{B}{l} \end{cases} \tag{6.29a}$$

- For a circle and square footing:

$$\begin{cases} S_c = 1 + \dfrac{N_q}{N_c} \\ S_q = 1 + \tan\varphi \\ S_\gamma = 0.60 \end{cases} \tag{6.29b}$$

As subjected to an eccentric load, for a strip footing, one should replace the width $B$ by $B' = B - 2e$ ($e$ is eccentricity); for a rectangle footing, replace the area $A$ by $A' = B'l'$, in which $B' = B - 2e_b$, $l' = l - 2e_l$, $e_b$ and $e_l$ are the eccentricities of load at short and long sides, respectively.

As the footing subjected the eccentric and inclined load, the bearing capacity is given by

$$p_u = cN_cS_ci_c + qN_qS_qi_q + \frac{1}{2}\gamma BN_\gamma S_r i_\gamma \tag{6.30}$$

in which $i_c$, $i_q$ and $i_\gamma$ are load inclination factors.

$$i_c = \begin{cases} 1 - \dfrac{mH}{B'l'cN_c} & (\varphi = 0) \\ i_q - \dfrac{1 - i_q}{N_c \tan\varphi} & (\varphi > 0) \end{cases}, \tag{6.31a}$$

$$i_q = \left(1 - \frac{H}{Q + B'l'c \cdot \cot\varphi}\right)^m, \tag{6.31b}$$

$$i_\gamma = \left(1 - \frac{H}{Q + B'l'c \cdot \cot\varphi}\right)^{m+1}, \tag{6.31c}$$

where $Q$ is vertical component of inclined loads (kN), $H$ is vertical component of inclined loads (kN), $B'$ is effective width of footing (m), $l'$ is effective width of footing (m), $m$ is constant.

For the strip footing, $m = 2$; as the load inclined to

- the short side of footing:

$$m_b = \frac{2 + (B/l)}{1 + (B/l)};$$

- the long side of footing:

$$m_l = \frac{2 + (l/B)}{1 + (l/B)};$$

and inclined arbitrary, $m_n = m_l \cos^2\theta_n + m_b \sin^2\theta_n$, $\theta_n$ is the inclined angle of load (°).

If the shear strength of soil above the footing base is considered, the bearing capacity is given by

$$p_u = cN_cS_ci_cd_c + qN_qS_qi_qd_q + \frac{1}{2}\gamma BN_\gamma S_\gamma i_\gamma d_\gamma, \tag{6.32}$$

where $d_c$, $d_q$ and $d_\gamma$ are the modified coefficient of embedment depth of footing

$$d_q = \begin{cases} 1 + 2\tan\varphi(1 - \sin\varphi)^2(d/B) & (d \leq B) \\ 1 + 2\tan\varphi(1 - \sin\varphi)^2 \tan^{-1}(d/B) & (d > B) \end{cases}, \tag{6.33a}$$

$$d_c = \begin{cases} 1 + 0.4d/B & (\varphi = 0, d \leq B) \\ 1 + 0.4\tan^{-1}(d/B) & (\varphi = 0, d > B) \\ d_q - \frac{1-d_q}{N_c \tan\varphi} & (\varphi > 0) \end{cases}, \tag{6.33b}$$

$$d_\gamma = 1. \tag{6.33c}$$

## Exercises

6.1 The width $b = 2.5\,\text{m}$ and the depth $d = 1.2\,\text{m}$ of a strip footing. The soil is the isotropic clay. $c = 12\,\text{kPa}$, $\varphi = 18°$, $\gamma_0 = 19.0\,\text{kN/m}^3$. Calculate $p_{cr}$ and $p_{\frac{1}{4}}$.

6.2 A strip footing with 1.50 m width is located at a depth of 1.0 m below the ground surface, The soil profile at the site is plain fill (with 0.8 m thickness, $\gamma_1 = 18\,\text{kN/m}^3$, water content is 35%) and clay (the thickness is 6 m, $\gamma_2 = 18.2\,\text{kN/m}^3$, water content is 38%, $c = 10\,\text{kPa}$, $\varphi = 13°$, $d_s = 2.72$). (1) Determine the critical edge pressure $p_{cr}$, $p_{\frac{1}{4}}$ and $p_{\frac{1}{3}}$; (2) assuming the groundwater level is 1.0 m below the ground surface and keeping the strength parameters unchanged, determine $p_{cr}$, $p_{\frac{1}{4}}$ and $p_{\frac{1}{3}}$.

6.3 A strip footing 1.50 m width is located at a depth of 1.20 m below the ground surface in the clay ($\gamma = 18.4\,\text{kN/m}^3$, $\gamma_{sat} = 18.8\,\text{kN/m}^3$, $c = 8\,\text{kPa}$, $\varphi = 15°$). According to the Terzaghi's bearing capacity, (1) determine the bearing capacity of soil with the general shear failure, and the allowable bearing capacity assuming a factor of safety of 2.5; (2) compare the bearing capacity when the footing is located at the depth of 1.6 m and 2.0 m, respectively.

## Bibliography

M. Budhu (2007). *Soil Mechanics and Foundation.* John Wiley & Sons, Inc.
R. F. Craig (2004). *Craig's Soil Mechanics* (Seventh edition). CRC Press.

E. E. DeBeer (1970). Experimental Determination of the Shape Factors and the Bearing Capacity Factors of Sand. *Geotechnique*, 20(4): 347–411.

L. Hongjian and L. Houxiang (2013). *Soil Mechanics* (in Chinese). Higher Education Press.

H. Liao (2018). *Soil Mechanics* (Third Edition). Higher Education Press, Beijing.

G. G. Meyerhof and T. Koumoto (1987). Inclination Factors for Bearing Capacity of Shallow Footings. *J. Geotech. Eng. Div. ASCE*, 113(9): 1013–1018.

A. W. Skempton (1951). *The Bearing Capacity of Clay*. Building Research Congress, London.

K. Terzaghi, R. B. Peck, and G. Mesri (1996). *Soil Mechanics in Engineering Practice* (Third Edition). John Wiley & Sons, Inc.

D. M. Wood (1990). *Soil Behavior and Critical State Soil Mechanics*. Cambridge University Press, Cambridge.

B. Zadroga (1994). Bearing Capacity of Shallow Foundations on Noncohesive Soils. *J. Geotech. Eng.*, 120(11): 1991–2008.

# Chapter 7

# Slope Stability Analysis

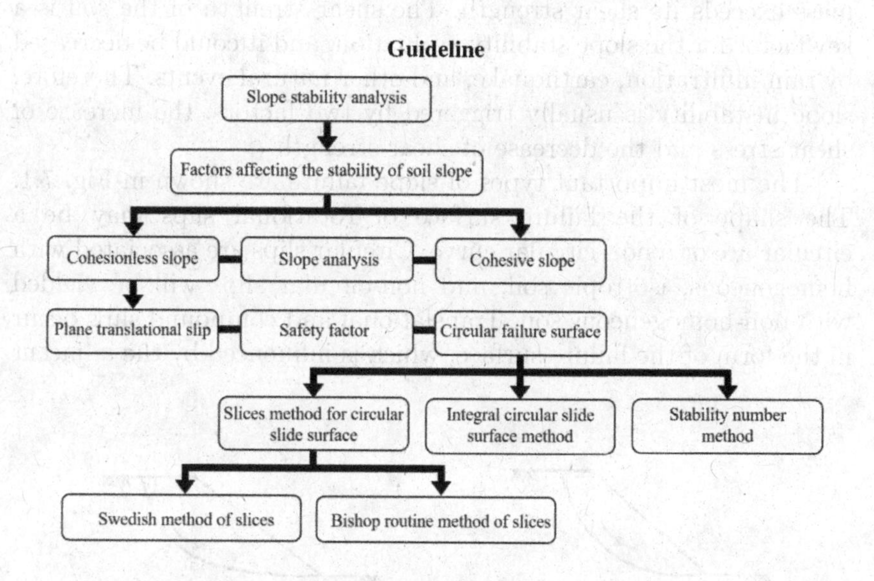

## 7.1 Introduction

Slope stability refers to the equilibrium state of inclined soil or rock slopes to slip or undergo movement along a particular failure surface. Slope stability analysis can be applied for earth-fill dams, embankments, excavated slopes, and natural slopes in soil and rock. It is a critical subject of study and research in soil mechanics, geotechnical engineering, and engineering geology. Slope instability usually happens for the natural soil slope, artificial earth-fill embankment, road cutting, and pit excavation. It usually happens due to the loading on the slope crest or the excavation in the slope foot, and the shear stress changes in the slope due to the gravity and seepage of the soil. The slope will easily slip when the shear stress of the soil mass exceeds its shear strength. The shear strength of the soil is a key factor for the slope stability evaluation, and it could be decreased by rain infiltration, earthquake, and other natural events. Therefore, slope instability is usually triggered by two factors: the increase of shear stress and the decrease of shear strength.

The most important types of slope failure are shown in Fig. 7.1. The shape of the failure surface of rotational slips may be a circular arc or a non-circular curve. Circular slips are associated with homogeneous, isotropic soil, and non-circular slips will be yielded with non-homogeneous soil. Translational and compound slips occur in the form of the failure surface, which is influenced by the adjacent

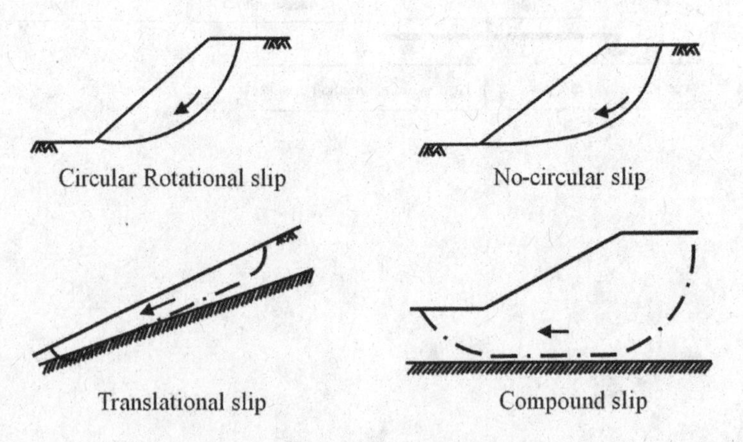


Fig. 7.1. Types of slope failure.

stratum with different shear strengths. Most of the failure surface is likely to pass through the soil stratum with lower shear strength. The form of the surface would also be influenced by discontinuities such as fissures and pre-existing slips. Translational slips will occur at a relatively shallow depth below the surface of the slope in the adjacent stratum and the failure surface tending to be plane and roughly parallel to the slope. Compound slips usually occur in the adjacent stratum with greater depth, and the failure surface consists of the curved and straight surfaces. In most cases, slope stability can be considered as a two-dimensional problem, i.e. the condition of plane strain.

The slope stability analysis can be divided into static and dynamic problems, and it can also be evaluated via analytical or empirical methods. Slope stability analysis is generally aimed at searching the critical state and the most dangerous failure surface for the slope failure. This chapter will mainly focus on introducing the main characteristics of slope and slide mass, as well as the commonly used calculation methods and engineering measures of soil slope stability analysis.

## 7.2 Factors and Engineering Measures that Influence Soil Slope Stability

Instability of soil slope is a feature that often occurs in actual engineering. The soil slope will slide if it is not correctly controlled, and it will have a significant impact on the progress of engineering. The slope failure will also cause accidents, thus threatening human lives and properties. The main influencing factors and the commonly faced problems in slope stability in engineering will be described briefly.

### 7.2.1 *Essential characteristics of slopes*

Slopes refer to rock and soil mass with an inclined surface. The slopes which are formed by natural geological processes, such as hill slopes or cliffs, are called natural slopes. The slopes formed by artificial excavation or backfilling, such as the slope of channel, foundation pit, embankment, are called artificial slopes. The components of the slope are shown in Fig. 7.2.

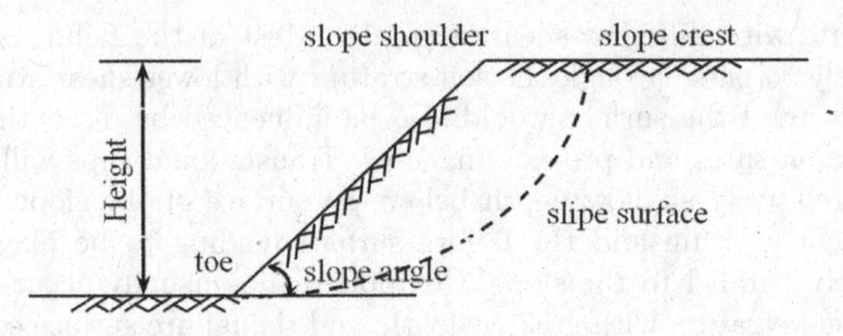


Fig. 7.2. Components of slope.

### 7.2.2 *Factors influencing for slope stability*

The factors that mainly influence slope stability are as follows:

(1) **Slope angle, $\beta$:** In general, slope stability increases with a decrease in the slope angle. It would be uneconomic in foundation pit excavation. A steep slope is economical for excavation but is not safe enough.
(2) **Slope height, $H$:** Similar to the slope angle $\beta$, the slope will have low safety with a large slope height.
(3) **The physical and mechanical properties of the soil** (e.g. unit weight $\gamma$, shear strength parameters $c$ and $\varphi$)**:** The slope is unstable due to the increase of the unit weight and shear stress of soil. The stable slope can transform to an unstable state due to several reasons, e.g. external loading, shear strength reduction, and pore water pressure generation, as a result of earthquakes, rainfall infiltration, and fluctuation of underground water level, respectively.
(4) **The infiltration of rainfall and seepage of underground water:** Rainfall infiltration could increase the water content of the soil, and the pore water in the soil plays a lubricating role in the weak layers of the slope. When the groundwater seepage is flowing in the slope, it could be safe for the slope to slide in a direction opposite to the direction of the seepage flow force; however, it is dangerous when both of them move along the same direction. The extensive practical engineering experience has also proved that landslides and slope failures often occur during the rainy season or are caused by heavy rain.
(5) **The influence of vibration:** The strength of soil can be decreased by the vibration load yield due to the pile compaction,

dynamic compaction, engineering blasting, vehicle movement, etc. On the contrary, soil strength also can be reduced as a result of earthquakes, e.g. the liquefaction of the soil in slope can be induced by the seismic load. The saturated, loose, and fine sand is more accessible to liquefaction due to the shock of the earthquake.

(6) **The impact of human activities and the ecological environment:** Slope stability could be significantly influenced by human activities, such as preloading on the slope crest, excavation, or river erosion at the toe of the slope. Climatic changes can also influence the slope stability, e.g. change the soil from dry to wet, shrink to expand, and freeze to melt. Ultimately, the soil will be softened and its strength will be reduced.

In conclusion, slope instability is usually triggered by the external influencing factors, e.g. the shear strength of soil can be reduced by the external influencing factors. Thus, additional attention should be paid to the impact of external influencing factors to enhance the stability and safety of the slope.

### 7.2.3 *Engineering design and measures*

In engineering practice, the stability and safety of the slope are evaluated by the factor of safety $K$, which indicates the safety of the slope under the most dangerous conditions. The value of $K$ is usually defined as the ratio of the slip resistant force (moment) to the sliding force (moment) on the failure surface. When the failure surface is fixed, the value of $K$ can be obtained via the method of slope stability analysis to determinate the slope stability.

In order to enhance the stability of the hill slope, cutting slope, and embankment, the factor of safety $K$ should be greater than 1.0. Taking into account the class of buildings, the condition of the foundation, the strength parameter of the soil, the average consolidation degree of the soil, the calculation method, and local experience, $K$ is determined to be 1.1–1.5. Some situations that are prone to slope instability in engineering are as follows:

(1) **Foundation pit excavation:** In general, the foundation depth of shallow clay foundation $d$ = 1–2 m. Vertical excavation can be used to save the earthwork, and the fast mechanized

construction can also be used. However, when the foundation depth is greater than 5 m, the foundation pit will collapse due to vertical excavation. If the slope of the foundation pit is too gentle, it will lead to an increase in the workload and the have an effect on the adjacent buildings. Thus, taking into account the above reasons, it is necessary to design a safe and economic slope of foundation pit through the slope stability analysis.

(2) **Loading on the slope crest:** Constructing the building or stacking heavy objects on the top of the slope may render the stable slope unstable and lead to sliding. If the building is far away from the slope, there is no influence on the slope. Therefore, a safe distance should be determined according to the engineering requirements.

(3) **The artificial river embankment, earth-filling earth dam, road embankment, and cutting slope:** A lot of time and work could be saved with an appropriate slope design for these engineering projects where the slope is at a very long distance.

## 7.3 Stability Analysis of Cohesionless Soil Slope

### 7.3.1 *Cohesionless soil slope stability analysis without seepage flow*

According to the actual observation, the failure surface of both the homogeneous or non-homogeneous sandy soil slope is similar to a plane. Therefore, in order to simplify the calculation, the plane failure surface method is often used for the slope stability analysis of sand, gravel, rockfill, and other cohesionless soil slopes.

Figure 7.3 shows a homogeneous cohesionless soil slope. Assuming that the top and bottom surfaces of the soil slope are horizontal and extend to infinity, the slope angle is $\beta$. The whole soil slope will be stable as long as the soil unit located on the slope is stable, irrespective of whether the soil is dry or wet. Regardless of the influence of stress on the stability of the unit body on both sides, the tangential force lets the soil unit of the slope surface slide down along the slope. Without considering the seepage, if the body force (gravity) of the soil unit is $W$, the sliding force is the component of $W$ along the slope only:

$$T = W \sin \beta. \tag{7.1}$$

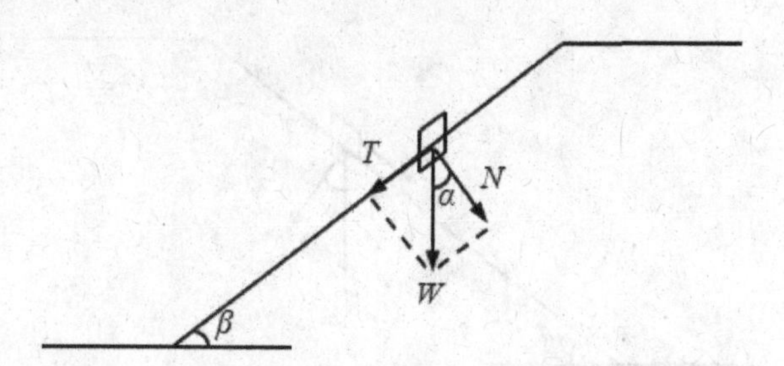


Fig. 7.3. Cohesionless soil slope without seepage flow.

The force that resists the sliding effect of the soil unit is the shear strength between the soil unit and the soil below. According to Coulomb's law, the sliding resistance force can be written as follows:

$$T_f = N \tan \varphi = W \cos \beta \tan \varphi, \tag{7.2}$$

where $N$ is the component of $W$ for the soil unit along the slope surface (kN) and $\varphi$ is the friction angle of the soil (°).

The safety factor of cohesionless soil slope can be defined as the maximum shearing resistance divided by the ratio of the tangential force:

$$K_s = \frac{T_f}{T} = \frac{W \cos \alpha \tan \varphi}{W \sin \beta} = \frac{\tan \varphi}{\tan \beta}. \tag{7.3}$$

Thus, the stability of the homogeneous cohesionless soil slope depends only on the slope angle and is independent of the slope height. The cohesionless soil slope is stable if the slope angle is lower than the friction angle of the soil. The slope is in the limit equilibrium state when $\beta = \varphi$ and $K_s = 1$. Therefore, the repose angle of the cohesionless soil slope is equal to the friction angle of the soil.

### 7.3.2 *Stability analysis of cohesionless soil slope with seepage flow*

The rapid drawdown of the water table will yield the seepage force in the soil of the slope surface, and it will render the cohesionless soil slope unstable. A soil unit is taken from the seepage overflow point on the slope, and both the body force $W$ and the seepage force $J$ act on the soil unit (see Fig. 7.4). The tangential force causing the soil

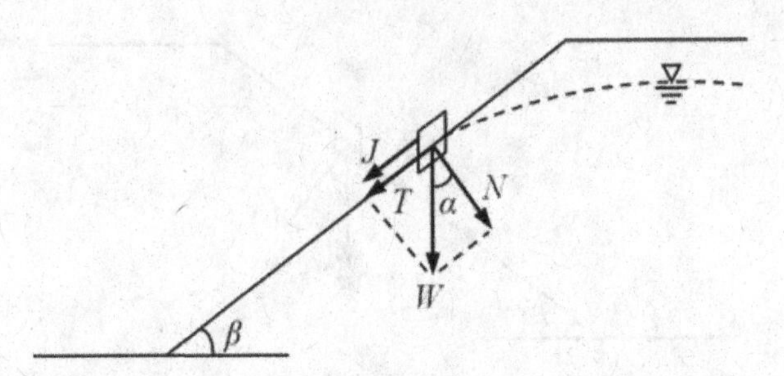


Fig. 7.4. Cohesionless soil slope under seepage flow.

sliding is as follows:

$$T + J = W \sin\beta + J. \tag{7.4}$$

The maximum shearing resistance force of the soil unit still is $T_f$, so the safety factor can be written as

$$K_s = \frac{T_f}{T + J} = \frac{W \cos\beta \tan\varphi}{W \sin\beta + J}. \tag{7.5}$$

For the unit soil mass, the effective unit weight of the soil is the floating weight $\gamma'$. The seepage force $J$ is equal to $\gamma_w i$, where $\gamma_w$ is the unit weight of water and $i$ is the hydraulic gradient of the seepage flow. Because the assumption is that the hydraulic gradient $i$ is approximately equal to $\sin\alpha$, Eq. (7.5) can be written as

$$K_s = \frac{\gamma' \cos\beta \tan\varphi}{(\gamma' + \gamma_w)\sin\beta} = \frac{\gamma' \tan\varphi}{\gamma_{\text{sat}} \tan\beta}, \tag{7.6}$$

where $\gamma_{\text{sat}}$ is the saturation weight of the soil (kN/m$^3$). Comparing this with the formula without seepage, Eq. (7.3), the difference is $\gamma'/\gamma_{\text{sat}}$, which is close to 1/2. Therefore, when the seepage is along the slope, the safety factor of the cohesionless soil slope will be reduced by half.

## 7.4 Stability Analysis of a Cohesive Soil Slope

Determining the slope of the failure surface is one of the key requirements in slope stability analysis. It has been proved that the cohesionless soil slope has a plane failure surface and the failure surface of a straight line in the cross-section view. However, the failure surface of the cohesive soil slope tends to be a curved surface, which is often approximated as a cylindrical surface in theoretical analysis. Thus, the safety factor of slope calculation can be simplified using the assumption of the circular arc failure surface. This assumption for cohesive soil slope stability analysis proposes a simplified and convenient way for safety factor calculation, i.e. the circular slip surface method.

The circular slip surface method was first proposed by Petterson (1916) and was then extensively studied and improved by Fellenius (1922) and Taylor (1937). The specific methods are the Swedish circular arc method, slice method, friction circle method, total stress method, effective stress method, and stability number method (Taylor, 1937), etc. These methods have different influencing factors and application conditions, but the same assumption of the circular arc slide surface and the limit equilibrium state to calculate the safety factor of the slope. Two types of methods can be summarized for those methods: the first one is the overall stability analysis method with a circular arc failure surface for slope stability analysis, which is mainly applicable to homogeneous and simple soil slope. The second is the slice method, and it can be applied to the soil slope with non-homogeneous soil, which have a complex structure and are submerged in water. This chapter mainly introduces the Swedish slice method with a circular slip surface, the Fellenius method for determining the failure surfaces of the most dangerous surfaces, the Bishop method of the circular slip surface, and the stability number method.

### 7.4.1 *Swedish slice method with circular slip surface*

The shear strength of the cohesive soil includes friction resistance and cohesion. The cohesive soil slope does not slide along the failure surface of the cohesionless soil slope. The most dangerous failure surface of the cohesive soil slope always cuts into the interior of the soil, thus yielding a circular arc failure surface.

#### 7.4.1.1 *The conception of the slice method*

The shear strength of each point on the failure surface is closely related to the normal stress at that point if the value of $\varphi$ is larger than zero. Normal stress is caused by the soil weight at each point on the failure surface. Thus, the shear strength at each point on the failure surface is also varied. In order to determine the magnitude of the normal stress of the failure surface (or the stress distribution on the failure surface), the commonly used method is to divide the sliding soil into several slices as a result of which the force on each slice can be easily analyzed. The formula of the safety factor can be established using the static force equilibrium conditions of each slice. This method is called the slicing method, which was proposed and improved by the Swedish railway engineers Petersen and Fellenius. It can be applied for the slope stability analysis problem with the circular slip surface or the non-circular slip surface and in some complex situations (such as seepage force or seismic force). The slice method is always applied to the plane strain problem (two-dimensional).

#### 7.4.1.2 *The basic ideas*

Figure 7.5(a) is assumed to be the cross-section of the soil slope drawn in a certain proportion, which represents a slope sliding along the circular arc failure surface. The sliding soil mass is divided into several slices, then the forces acting on the slice include the body force of the slice $W_i$, the normal force and tangential resistance of the bottom surface of the slice $\bar{N}_i$ and $\bar{T}_i$, and the horizontal and vertical forces acting on both sides of the slice $E_i$, $X_i$ and $E_{i+1}$, $X_{i+1}$. Slice $i$ was taken out for force equilibrium analysis, as shown in Fig. 7.5(b).

If the number of slices is large enough and the width of each slice is small enough. It can be assumed that $\bar{N}_i$ acts on the midpoint of

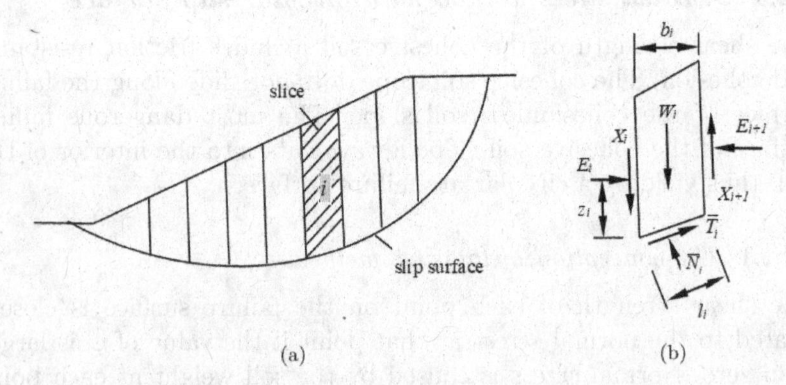


Fig. 7.5. The slice and the force equilibrium conditions.

the bottom surface of the slice, and the resultant forces of the forces acting on the two sides of slice $i$, i.e. the normal component forces $X_i$, $X_{i+1}$ and the tangential forces $E_i$, $E_{i+1}$, can be approximately considered to be equal in magnitude and opposite in direction. According to Coulomb's law, the shear strength of the failure surface should be

$$\tau_{fi} = c_i + \sigma_i \tan \varphi_i = \frac{c_i l_i + \bar{N}_i \tan \varphi_i}{l_i}, \tag{7.7}$$

where $l_i$ is the arc length (m) of the failure surface of the slice $i$, $c_i$ is the cohesion (kPa) of the soil on the failure surface of the slice $i$, and $\varphi_i$ is the friction angle (°) of the soil on the failure surface of the slice $i$. According to the definition, the formula for the safety factor (see Eq. (7.7)) can be written as follows:

$$K_s = \frac{T_f}{\bar{T}_i} = \frac{\tau_{fi} l_i}{\bar{T}_i} = \frac{c_i l_i + \bar{N}_i \tan \varphi_i}{\bar{T}_i}. \tag{7.8}$$

Therefore, the relation between $\bar{T}_i$ and $\bar{N}_i$ can be figured out as

$$\bar{T}_i = \frac{\tau_{fi} l_i}{K_s} = \frac{c_i l_i + \bar{N}_i \tan \varphi_i}{K_s}. \tag{7.9}$$

The number of slices is $n$, and the number of unknown quantities is shown in Table 7.1. There are only two static equilibrium equations for forces, only one equation for moments, and the total number of equations is $3n$, so the number of unknowns $(n - 2)$ cannot be solved. Therefore, the general soil slope stability analysis is always a statically indeterminate problem. In order to transform it into a statically determinate problem, it is necessary to simplify the assumption of the force acting on the slices.

#### 7.4.1.3 *Basic assumptions*

The Swedish slice method of circular slip surface is assumed as follows:

(1) The failure surface is a cylinder and the sliding soil mass is a rigid body (no deformations).
(2) The forces acting on both sides of the slice are not considered.

Table 7.1. Unknown quantities and the corresponding number of the circular arc slip surfaces.

| Unknown quantities | Number |
|---|---|
| Safety factor, $K_s$ | 1 |
| Normal force of slice bottom, $\bar{N}_i$ | $n$ |
| Normal force between the slices, $E_i$ | $n-1$ |
| Shear force between the slices, $X_i$ | $n-1$ |
| Location of the force between the slices, $z_i$ | $n-1$ |
| Total | $4n-2$ |

According to the above assumptions, the unknowns quantities in Table 7.1 are reduced to $(3n-3)$. We thus have only $(n+1)$ unknown quantities. The value of $K_s$ and $\bar{N}_i$ can be obtained by the force equilibrium equation of each slice and the moment equilibrium equation of the whole sliding soil mass.

#### 7.4.1.4 *Derivation of a formula for the Swedish slice method with a circular slip surface*

A homogeneous soil slope is used (shown in Fig. 7.6(a)) for deriving the formula of the Swedish slice method with a circular slip surface.

A proportional slope cross-sectional is shown in Fig. 7.6(a). $AC$ is a circular slip surface and the center of the circle is $O$, the radius is $R$, and sliding soil mass $ABC$ is a rigid block that slides around the circle center $O$ along the $AC$ surface. The sliding soil mass $ABC$ can be divided into several slices with uniform width. Generally speaking, the smaller the width of the slice is, the higher the calculation accuracy will be. However, in order to simplify the calculation and meet the design requirements, the width of the slice is selected to be 2–6 m or $0.1R$ is usually used.

As shown in Fig. 7.6(b), a slice $i$ was selected to analyze the force equilibrium. According to the basic assumption of the Swedish slice method with a circular slip surface, the forces acting on the slice $i$ are as follows: the body force of the slice $W_i$, the normal reaction $\bar{N}_i$, and the tangential reaction $\bar{T}_i$ acting on the bottom surface of the slice. These forces are discussed as follows:

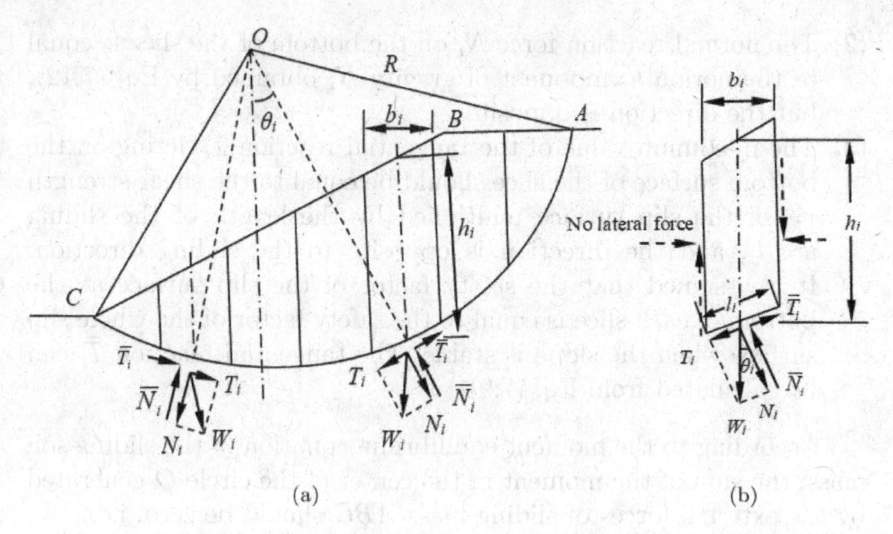


Fig. 7.6. Calculation scheme of the Swedish method.

(1) The gravity of the slice $W_i$ is vertically downward, and its value is

$$W_i = \gamma b_i h_i, \tag{7.10}$$

where $\gamma$ is the unit weight of the soil (kN/m$^3$) and $b_i$ and $h_i$ are the width and average height of the slice (m), respectively.

Let the unit weight be $W_i$ along its centroid action line to the failure surface of the slice, and it can be divided into the normal force $N_i$ passing through the center of the circular slip surface and the tangential force $T_i$ acting on the circular slip surface. If $\theta_i$ represents the intersection angle between the normal and the vertical line at the midpoint of the bottom surface of the slice, the body force of the slice in the normal direction and the tangent direction is as follows:

$$N_i = W_i \cos\theta_i, \tag{7.11}$$

$$T_i = W_i \sin\theta_i, \tag{7.12}$$

where $T_i$ is the force pushing the soil down. However, as shown in Fig. 7.6(b), if the slice $i$ is located to the left of the vertical line through the center of the slip arc, then $T_i$ is the slip resistance. The value of slip resistance $T_i$ is very small, which is conducive to stability, so it can be neglected.

(2) The normal reaction force $\bar{N}_i$ on the bottom of the slice is equal to the normal component of gravity $N_i$ obtained by Eq. (7.12), but the direction is opposite.
(3) The maximum value of the tangential reaction $\bar{T}_i$ acting on the bottom surface of the slice should be equal to the shear strength $\tau_{fi}$ of the slip surface multiplied by the length of the sliding arc $l_i$, and the direction is opposite to the sliding direction. It is assumed that the safety factor of the slip surface at the bottom of each slice is equal to the safety factor of the whole slip surface when the slope is stable. The tangential reaction $\bar{T}_i$ can be calculated from Eq. (7.9).

According to the moment equilibrium equation of the sliding soil mass, the sum of the moment in the center of the circle $O$ generated by the external forces of sliding mass $ABC$ should be zero, i.e.

$$\sum M_{oi} = 0. \tag{7.13}$$

The sum of the sliding moment generated by the body force $W_i$ of each slice is

$$\sum M_{si} = \sum W_i R \sin\theta_i = \sum \gamma b_i h_i R \sin\theta_i. \tag{7.14}$$

The normal reaction on the circular slip surface passes through the center of the circle $O$, and no moment will be yielded. According to Eq. (7.9), the slip resistant moment generated by the tangential reaction $\bar{T}_i$ on the circular slip surface is

$$\begin{aligned}\sum M_{ri} = \sum \bar{T}_i R &= \frac{\sum (cl_i + \bar{N}_i \tan\varphi)}{K_s} R \\ &= \frac{\sum (cl_i + W_i \cos\theta_i \tan\varphi)}{K_s} R \\ &= \frac{\sum (cl_i + \gamma b_i h_i \cos\theta_i \tan\varphi)}{K_s} R.\end{aligned} \tag{7.15}$$

If the sliding moment and the anti-sliding moment are in equilibrium and Eqs. (7.14) and (7.15) are equivalent, then

$$\sum \gamma b_i h_i R \sin\theta_i = \frac{\sum(cl_i + \gamma b_i h_i \cos\theta_i \tan\varphi)}{K_s} R. \tag{7.16}$$

Thus, it can be concluded that

$$K_s = \frac{\sum (cl_i + \gamma b_i h_i \cos\theta_i \tan\varphi)}{\sum \gamma b_i h_i \sin\theta_i}. \tag{7.17}$$

If the width of each slice is uniform, Eq. (7.17) can be simplified as follows:

$$K_s = \frac{c\widehat{L} + \gamma b \tan\varphi \sum h_i \cos\theta_i}{\gamma b \sum h_i \sin\theta_i}, \tag{7.18}$$

where $\widehat{L}$ is the arc length of arc $AC$.

Moreover, we should pay attention to the position of each slice. If the center of the bottom surface of the slice is on the right of the vertical line through the center of the circular slip surface $O$, the direction of the tangential force $T_i$ is in the same direction of the sliding soil mass, and the sign of $T_i$ should be positive. However, if the center of the bottom surface of the slice is to the left of the vertical line through $O$ and the direction of tangential force $T_i$ is opposite to the sliding direction and plays a slip resistant role, then the sign of $T_i$ should be negative (see Fig. 7.6(a)). The tangential reaction $\bar{T}_i$ must be opposite to the sliding direction.

Generally speaking, the most popular calculation methods for the slope stability analysis are always used to calculate the $K_s$ value of different safety factors by assuming many different circular slip surfaces and obtaining the safety factor of the most dangerous slip surface (the minimum safety factor $K_{\mathrm{smin}}$ value), and this safety factor is the final solution of this slope stability problem. If the solution of the safety factor is failing to meet the design requirements, the slope stability analysis must proceed again via the design adjustment or reinforcement. $K_{\mathrm{smin}}$ should be greater than 1.25 for engineering applications. If there is pore water pressure in the slope soil, the Swedish slice method can be also expressed by the effective stress method for slope stability analysis. The sliding moment and anti-sliding moment in Eqs. (7.14) and (7.15) can be expressed as follows:

$$\sum M_{si} = \sum W_i R \sin\theta_i = \sum \gamma' b_i h_i R \sin\theta_i, \tag{7.19}$$

$$\sum M_{ri} = \frac{\sum [c' l_i + (\gamma' b_i h_i \cos\theta_i - u_i l_i) \tan\varphi')}{K_s} R. \tag{7.20}$$

Therefore,

$$K_s = \frac{\sum [c'l_i + (\gamma' b_i h_i \cos\theta_i - u_i l_i)\tan\varphi']}{\sum \gamma' b_i h_i \sin\theta_i}, \tag{7.21}$$

where $c'$ is the effective cohesion of soil (kPa), $\varphi'$ is the effective friction angle of the soil (°), $\gamma'$ is the buoyancy unit weight of the soil (kN/m$^3$), and $u_i$ is pore water stress at the midpoint of the bottom surface of the slice $i$.

When the slope is subjected to steady seepage, the slip resistant effect of seepage force can be neglected and the formula of the safety factor calculation is as follows:

$$K_s = \frac{c'\widehat{L} + \Sigma(\gamma h_{1i} + \gamma' h_{2i} + \gamma' h_{3i}) b \cos\theta_i \tan\varphi'}{\Sigma(\gamma h_{1i} + \gamma' h_{2i} + \gamma' h_{3i}) b \sin\theta_i}, \tag{7.22}$$

where $\gamma_{\mathrm{sat}}$ is the saturation weight of the soil (kN/m$^3$) and $h_{1i}$, $h_{2i}$, and $h_{3i}$ are the distance between the saturation line and the top surface of the slice, the distance between the saturation line and the water level outside the slope, and the distance between the water level outside the slope and the bottom surface of the slice (m), respectively.

At present, the Swedish slice method with the circular slip surface is still widely used in engineering. The internal force between the soil slice is neglected in the Swedish slice method, the static equilibrium condition could not be fully satisfied, and the the overall equilibrium condition could only be satisfied.

### 7.4.2 *Fellenius method for the most dangerous slip surface determination*

Several slip surfaces must be assumed in the Swedish slice method, and the minimum safety factor corresponding to the most dangerous slip surface must be obtained via trial analysis. Therefore, the work of analysis is complicated and tedious, especially for the determination of the center position for the most dangerous slip surface. Fellenius proposed an empirical and simplified method for the most dangerous slip surface determination of the simple slope. Fellenius method is also a generally used method for the problem of slope stability analysis.

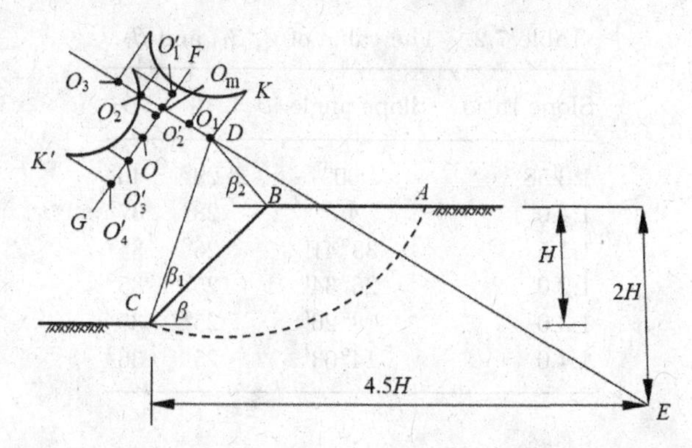


Fig. 7.7. Determination of the center position for the most dangerous slip surface.

(1) When the friction angle of the soil $\varphi = 0$, Fellenius proposed that the most dangerous slip surface of the slope passes through the toe of the slope, and the circle center point of the circular slip surface $D$ can be obtained by the intersection point using $\beta$, $\beta_1$, and $\beta_2$ (see Fig. 7.7).

(2) When the friction angle of the soil $\varphi > 0$, Fellenius proposed that the most dangerous slip surface of the slope also passes through the toe of the slope, and the center point of the circular slip surface $O$ is located on the extension line of $ED$ (see Fig. 7.7). The center point of the circular slip surface is upward with the value of friction angle $\varphi$. The position of point $E$ is determined by $2H$ vertical distance from the toe of the slope and $4.5H$ horizontal distance to the right. Numbers of the center point of the circular slip surface $O_1$, $O_2$, and $O_3$, etc., can be obtained from the extension line of $ED$. The safety factors $K_1$, $K_2$, and $K_3$, etc., can be obtained via the circular slip surfaces through the toe of the slope $C$. The safety factor $K$ value curve can be connected through $K_1$, $K_2$, and $K_3$, etc., and the lowest point of the $K$ line $O_m$ is the center point of the circular slip surface corresponding to the minimum safety factor. However, the center of the most dangerous slip surface may not appear on the $ED$ line. Some test circle center point $O_1'$, $O_2'$, and $O_3'$, etc., can be taken on the $FG$ line (the vertical line of $ED$ though $O_m$ point). Thus, the corresponding safety factors $K_1'$, $K_2'$, $K_3'$, ..., can be obtained. And then, the safety factor $K'$value curve can

Table 7.2. The value of $\beta$, $\beta_1$, and $\beta_2$.

| Slope ratio | Slope angle, $\beta$ | $\beta_1$ | $\beta_2$ |
|---|---|---|---|
| 1:0.58 | 60° | 29° | 40° |
| 1:1.0 | 45° | 28° | 37° |
| 1:1.5 | 33°41′ | 26° | 35° |
| 1:2.0 | 26°34′ | 25° | 35° |
| 1:3.0 | 18°26′ | 25° | 35° |
| 1:4.0 | 14°03′ | 25° | 36° |

be established, and the circle center point of the most dangerous slip surface $O$ corresponding to the minimum safety factor of $K'$ line can be fixed (Table 7.2).

It can be seen that the trial calculation work can be reduced via the semi-analytical and semi-graphical method proposed by Fellenius. But it is only limited to the work of determining the center point of the circle of the most dangerous slip surface.

### 7.4.3 *Bishop method with a circular slip surface*

Considering the action of the lateral forces of the slices, Bishop (1954) proposed a new method of slices for slope stability analysis. Taking the lateral forces ($E_i$, $X_i$, and $E_{i+1}$, $X_{i+1}$) on both sides of the soil slice into account, the safety factor for slip resistance on the slip surface of each soil slice equals the average safety factor of the whole slip surface. The basic ideas and the derivation of the formula for Bishop's method are as follows.

The cross-section of a soil slope is shown in Fig. 7.8(a), which shows a circular slip surface with center $O$ and radius $R$. The force acting on the slice $i$ is as follows: the gravity $W_i$, the tangential reaction force acting on the slip surface $\bar{T}_i$, the effective normal reaction force acting on the slip surface $\bar{N}'_i$, and pore water pressure $u_i l_i$. It is assumed that the action points of these forces are at the midpoint of the circular slip surface of the slice $i$.

The normal forces $(E_i, E_{i+1})$ and tangential force $(X_i, X_{i+1})$ acting on both sides of the slice $i$ are also considered in the Bishop's method, and $\Delta X_i = X_{i+1} - X_i$.

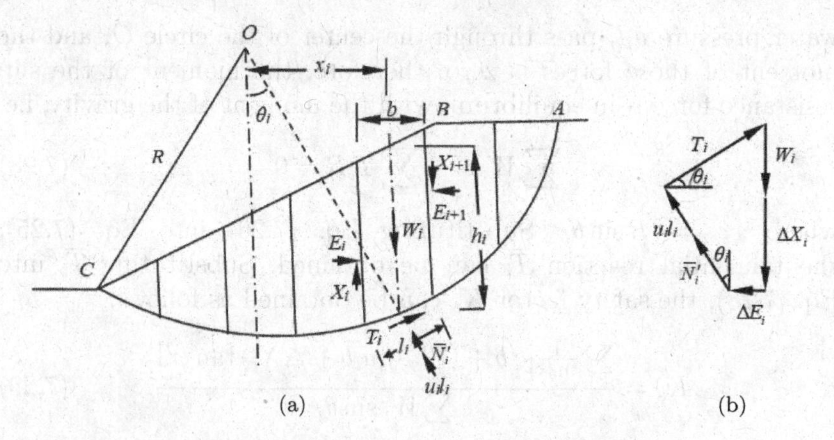


Fig. 7.8. Calculation scheme of Bishop method.

Take the equilibrium of the resultant force in the vertical direction of the slice $i$ (shown in Fig. 7.8(b)),

$$W_i + \Delta X_i - \bar{T}_i \sin\theta_i - \bar{N}'_i \cos\theta_i - u_i l_i \cos\theta_i = 0, \tag{7.23}$$

$$\bar{N}'_i \cos\theta_i = W_i + \Delta X_i - \bar{T}_i \sin\theta_i - u_i b. \tag{7.24}$$

The slope cannot slide if the sliding force is lower than the shear strength of the slip surface of the slice. Using the effective stress, the tangential reaction force of the slip surface of the slice can be expressed as follows:

$$\bar{T}_i = \frac{\tau_{fi} l_i}{K_s} = \frac{c' l_i}{K_s} + \bar{N}'_i \frac{\tan\varphi'}{K_s}. \tag{7.25}$$

Substituting Eq. (7.25) into Eq. (7.24), the effective normal reaction force can be obtained as follows:

$$\bar{N}'_i = \frac{1}{m_{\theta_i}}\left(W_i + \Delta X_i - u_i b - \frac{c' l_i}{K_s}\sin\theta_i\right), \tag{7.26}$$

where

$$m_{\theta_i} = \cos\theta_i + \frac{\tan\varphi'}{K_s}\sin\theta_i = \cos\theta_i\left(1 + \frac{\tan\varphi'\tan\theta_i}{K_s}\right). \tag{7.27}$$

The moment of the whole slide mass to the circle center $O$ is in equilibrium, so the moment of each side of the slice will offset each other. The action lines of the effective reaction force $\bar{N}'_i$ and pore water pressure $u_i l_i$ pass through the center of the circle $O$, and the moment of those forces is zero; therefore, the moment of the slip resistance force is in equilibrium with the moment of the gravity, i.e.

$$\sum W_i x_i - \sum \bar{T}_i R = 0 \tag{7.28}$$

where $x_i = R\sin\theta_i$. Substituting Eq. (7.26) into Eq. (7.25), the tangential reaction $\bar{T}_i$ can be obtained. Substituting $\overline{T_i}$ into Eq. (7.28), the safety factor $K_s$ can be obtained as follows:

$$K_s = \frac{\sum \frac{1}{m_{\theta_i}}[c'b + (W_i - u_i b + \Delta X_i)\tan\varphi']}{\sum W_i \sin\theta_i}. \tag{7.29}$$

This formula is Bishop's general equation for the factor of safety calculation, but $\Delta X_i$ is still unknown. In order to calculate $K_s$, the value of $\Delta X_i$ must be estimated, and it can be solved via the successive approximation method. The value of $X_i$ and $E_i$ shall satisfy the equilibrium condition of each slice, and $\sum \Delta X_i$ and $\sum \Delta E_i$ of the sliding soil mass are equal to zero. Bishop has proved that the error is only 1% if the value of $\Delta X_i$ is equal to zero. Thus, Eq. (7.29) can be simplified as follows:

$$K_s = \frac{\sum \frac{1}{m_{\theta_i}}[c'b + (W_i - u_i b)\tan\varphi']}{\sum W_i \sin\theta_i}. \tag{7.30}$$

This formula is Bishop's simplified equation for the factor of safety calculation. The pore pressure $u_i$ can be obtained by $\gamma h_i \bar{B}$, where $\bar{B}$ is the pore stress coefficient. Because the formula of $m_{\theta_i}$ also contains $Ks$, the trial calculation must be processed for $Ks$ calculation. First, we can assume that $Ks$ is equal to 1.0, and then $m_{\theta_i}$ can be calculated by Eq. (7.27), and then $Ks$ can be calculated by Eq. (7.30). If the value of $Ks$ is not equal to 1.0, the value of $Ks$ can be used to calculate the new value of $Ks$. The accuracy requirements can be met after three or four iterations for engineering application purposes, and the iteration is always convergent.

It must be noted that for those slices with negative $\theta_i$, the value of $m_{\theta_i}$ could approach zero. If this is the case, then Bishop's simplified method cannot be used. According to the suggestions of some researchers, the value of $m_{\theta_i}$ is equal to or lower than 0.2, thereby yielding a large error for calculating the value of $K_s$. $\bar{N}_i'$ will have a negative value when the value of $\theta_i$ of the slice is very large, and $\bar{N}_i'$ should be equal to zero in this situation.

In the case of seepage flow, the pore water pressure will act on both sides of the slice. Equation (7.22) is established via the Swedish method, so the force on both sides of the slice is not taken into account, the influence of seepage force on the sliding resultant is also not taken into account. Equation (7.22) can be rewritten via the Bishop method as follows:

$$K_s = \frac{\Sigma \frac{1}{m_\theta}\{c'b + [W_i - (u_i - \gamma_w z_i)b] \tan \varphi'\}}{\Sigma W_i \sin \theta_i}, \tag{7.31}$$

where $z_i$ is the distance between the water table above the middle point of the slice bottom ($m$). No seepage force term has appeared in Eq. (7.31), but the influence of the seepage force is reflected by the combination of the total weight of the soil (saturation unit weight) and the pore water pressure.

Simplified Bishop method assumes that $\Delta X_i$ is equal to zero, and $n - 1$ unknown quantities are reduced. The value of the safety factor $K_s$ can be obtained without calculating the magnitude and location of $E_i$ (the vertical forces and the moments of equilibrium of the sliding soil mass). Moreover, the simplified Bishop method also can't meet all the equilibrium conditions, so it is still not a strict method, and the error of the solution is about 2–7%.

## 7.5 Stability Number Method

A simplified chart method was proposed by Taylor (1948) for soil slope stability analysis and this method is much more efficient for the safety factor calculation of the homogeneous soil slopes with a height lower than 10 m. The stability number $N_s$ is a key factor for Taylor's stability chart method, and it can be defined as follows:

$$N_s = \frac{\gamma H}{c}. \tag{7.32}$$

The slope stability chart developed by Taylor (1937, 1948) is shown in Fig. 7.9. The $y$-coordinate is the stability number $N_s$, and the $x$-coordinate is the angle of the slope $\beta$. Two fundamental problems for simple slope stability analysis can be rapidly solved using Taylor's stability chart.

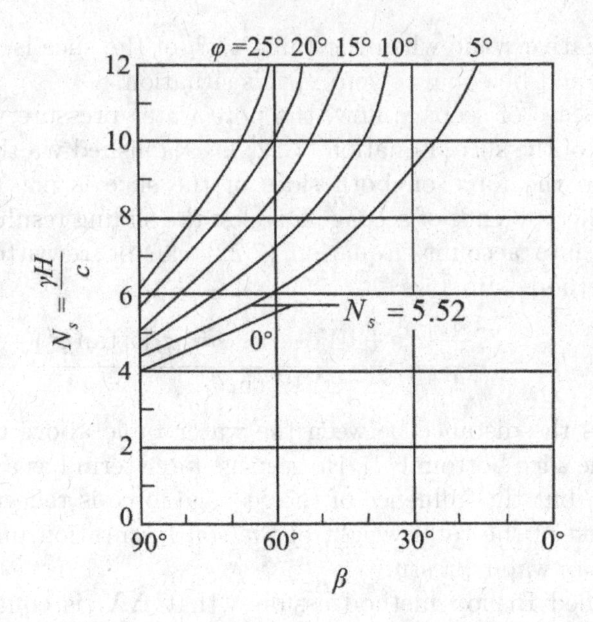


Fig. 7.9. Taylor's stability number chart (Taylor, 1937).

(1) The critical safety height of the slope $H_{\rm cr}$ will be obtained if the value of $\beta$, $\varphi$, $c$, and $\gamma$ are fixed, i.e. the value of $N_s$ can be obtained using the chart with $\beta$ and $\varphi$, and then, the value of $H_{cr}$ can be calculated with $c$ and $\gamma$.
(2) The stable slope angle of soil slope will be obtained if the value of $c$, $\varphi$, $\gamma$, and $H$ are fixed, i.e. the value of $N_s$ can be calculated with $c$, $\gamma$, and $H$, and then, the value of $\beta$ can be obtained using the chart with $N_s$ and $\varphi$.

Taylor obtained a $\beta$–$N_s$ relationship diagram for undrained clay ($\varphi = 0$). $\beta$–$N_s$ relationship is shown in Fig. 7.10.

A horizontal hard stratum located in the depth of $H_1 = n_d H$ below the upper ground surface. The stability number $N_s$ increases as the value of $n_d = H_1/H$ decreases. For $\beta > 53°$, the critical circle is a toe circle (see Fig. 7.11(a)) and the depth of the hard stratum $H_1$ has no effect on the stability number $N_s$. For $\beta < 53°$, the failure surface of slope depends on both the slope angle $\beta$ and the depth of the hard stratum $H_1$. For $n_d > 4.0$, $N_s = 5.52$ ($n_d$ can be assumed to be infinite), and $N_s$ is independent with the slope angle $\beta$, i.e.

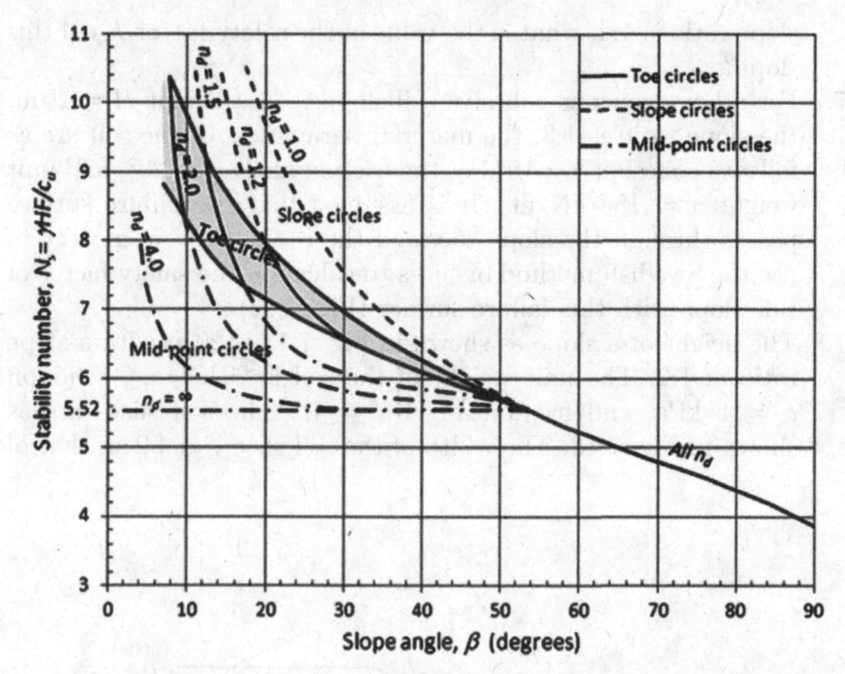


Fig. 7.10. Relationship between stability number and slope angle (Taylor, 1937).

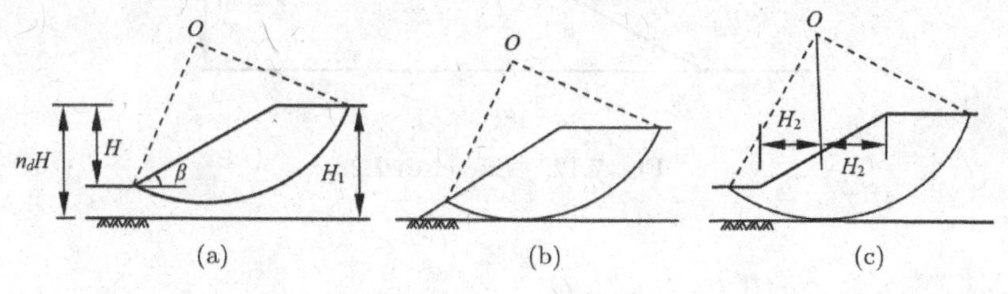


Fig. 7.11. Three types of failure surface. (a) Toe circle, (b) slope circle, and (c) mid-point circle.

the critical safety height of the slope $H_{\mathrm{cr}} = 5.52 \times c_u\gamma$, $c_u$ is the undrained shear strength of the soil.

## Exercises

7.1 The material parameters of a sand slope are as follows: saturated unit weight $\gamma_{\mathrm{sat}} = 19.0\,\mathrm{kN/m^3}$, friction angle $\varphi = 32°$, and slope ratio is 1:3 (height/run). Try to calculate the safety factor $K_s$ of this slope for dry and submerged conditions, respectively. If the slope ratio is 1:4, what is the value of the safety factor $K_s$ of this slope?

7.2 For a homogeneous cohesive soil slope with a height $H = 20\,\text{m}$, the slope ratio is 1:3, the material parameters of the soil are as follows: cohesion $c = 10\,\text{kPa}$, the friction angle $\varphi = 20°$, and unit weight $\gamma = 18.0\,\text{kN/m}^3$. It is assumed that the failure surface passes through the slope toe, and the radius $R = 55\,\text{m}$. Try to use the Swedish method of slices to calculate the safety factor of this slope with this failure surface (Fig. 7.12).

7.3 The height of a slope as shown in Fig. 7.13 is 20 m with a slope ratio of 1:2. The unit weight of the soil is $20\,\text{kN/m}^3$, cohesion $c = 45\,\text{kPa}$, and friction angle $\varphi = 7°$. The soil slices are as shown in Fig. 7.13. The width of the soil slice 7 is 4.2 m, that of

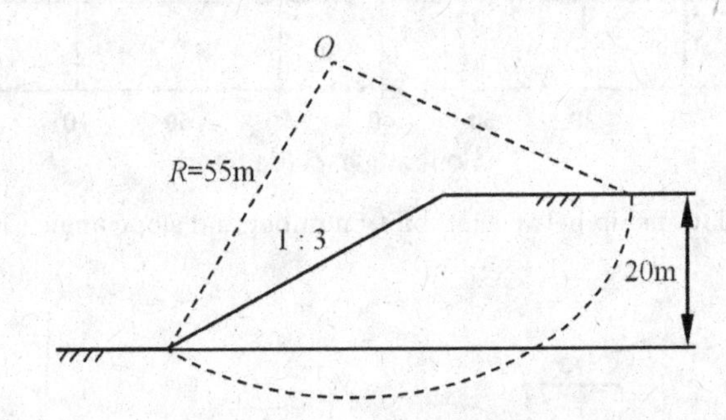


Fig. 7.12. Exercises 7.2.

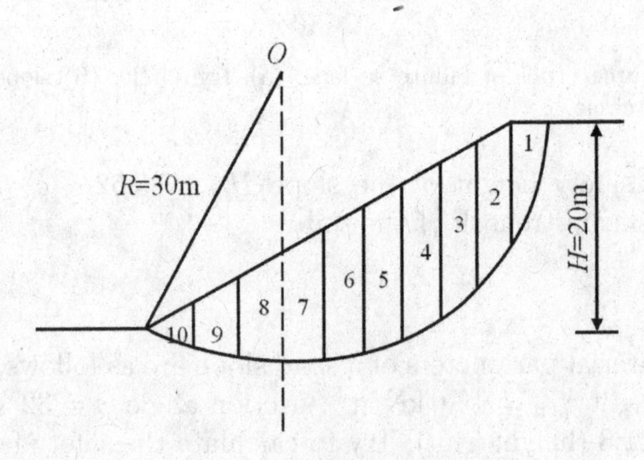


Fig. 7.13. Exercises 7.3.

soil slice 8 is 4.9 m, and that of the others is 4.5 m. Please analyze the stability of this slope using the Swedish slice method.

7.4 The slope angle of a uniform soil slope $\beta = 30°$, and the material parameters of the soil are as follows: $\varphi = 20°$, $c = 5\,\text{kPa}$, and $\gamma = 16\,\text{kN/m}^3$. Determine the critical safety height $H_{cr}$ of the soil slope.

7.5 For a soil slope of height $H = 10\,\text{m}$, the material parameters of the soil are as follows: $\varphi = 20°$, $c = 7.0\,\text{kPa}$, and $\gamma = 16.0\,\text{kN/m}^3$. Try to calculate the critical safety slope angle.

## Bibliography

A. W. Bishop (1955). The use of the slip circle in the stability analysis of slopes. *Geotechnique*, 5(1): 7–17.

A. W. Bishop and N. Morgenstern (1960). Stability coefficients for earth slopes. *Geotechnique*, 10(4): 129–150.

H. Liao (2018). *Soil Mechanics* (Third Edition). Higher Education Press, Beijing.

N. Janbu (1954). *Stability Analysis of Slopes with Dimensionless Parameters*, Harvard Soil Mechanics Series 46, Harvard University Press, Cambridge.

N. Janbu (1973). *Slope Stability Computations: Embankment Dam Engineering — Casagrande.* John Wiley and Sons, Inc., New York.

N. Morgenstern (1963). Stability charts for earth slopes during rapid drawdown. *Geotechnique*, 13(2): 121–131.

E. Spencer (1967). A method of analysis of the stability of embankments assuming parallel inter-slice forces. *Geotechnique*, 17(1): 11–26.

D. W. Taylor (1937). Stability of earth slopes. *J. Boston Soc. Civil Eng.*, 24: 197–247.

D. W. Taylor (1948). *Fundamentals of Soil Mechanics.* John Wiley and Sons, Inc., New York.

K. Terzaghi and R. B. Peck (1967). *Soil Mechanics in Engineering Practice.* John Wiley and Sons, Inc., New York.

Chapter 8

# Earth Pressure and Retaining Walls

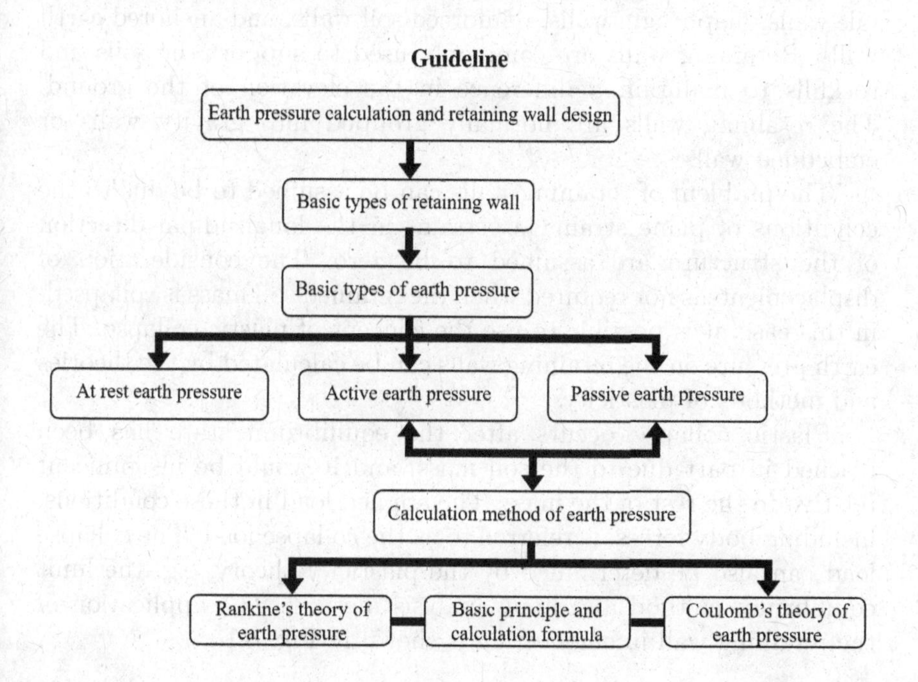

## 8.1 Introduction

This chapter mainly discusses the magnitude and distribution of lateral pressure between the soil and an adjoining retaining wall. The retaining wall is a structure that retains the geomaterial (e.g. soil or rock) and prevents the collapse or erosion of the geomaterial. The retaining wall is designed to resist the pressure of the geomaterial that it is holding back. The pressure (lateral pressure) exerted by the rock and the soil on the back surface of the wall is named the earth pressure, which can be calculated via the theory of soil mechanics, elasticity, and plasticity. Various types of retaining walls are used in civil engineering, with the general ones being: mass construction gravity walls, reinforced concrete walls, crib walls, gabion walls, Sheet pile walls, diaphragm walls, reinforced soil walls, and anchored earth walls. Retaining walls are commonly used to support the soils and rockfills to maintain a difference in the elevation of the ground. The retaining walls are normally grouped into gravity walls or embedded walls.

The problem of retaining walls can be assumed to be due to the conditions of plane strain, i.e. strains in the longitudinal direction of the structure are assumed to be zero. The consideration of displacements is not required when the retained soil mass is collapsed; in this case, it is possible to use the concept of plastic collapse. The earth pressure on the retaining walls can be calculated by the theories and methods of plasticity.

Plastic collapse occurs after the equilibrium state has been reached in part due to the soil mass, and it would be insignificant relative to the rest of the mass. The applied load in these conditions, including body forces, is referred to as the collapse load. The collapse load can also be determined by the plasticity theory, e.g. the limit equilibrium method and limit analysis method. The application of the retaining wall in engineering is shown in Fig. 8.1.

## 8.2 Earth Pressure on the Retaining Wall

$AA'$ is the top width of the wall, and $OB$ is the width of the wall bottom (shown in Fig. 8.2(a)). The distance $H$ from the top of the wall to the bottom of the wall is the vertical height of the wall. The surface that is directly in contact with the soil mass is named

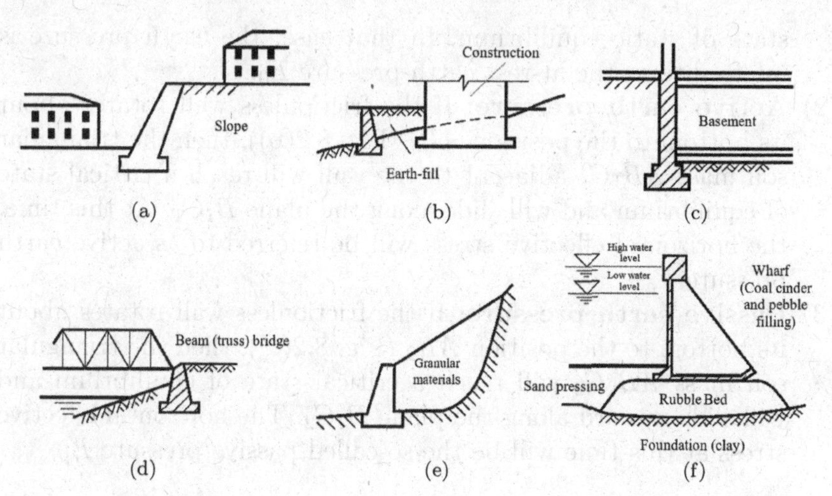


Fig. 8.1. Application of retaining wall in engineering. (a) Slope, (b) building, (c) basement, (d) bridge, (e) granular materials, and (f) wharf.

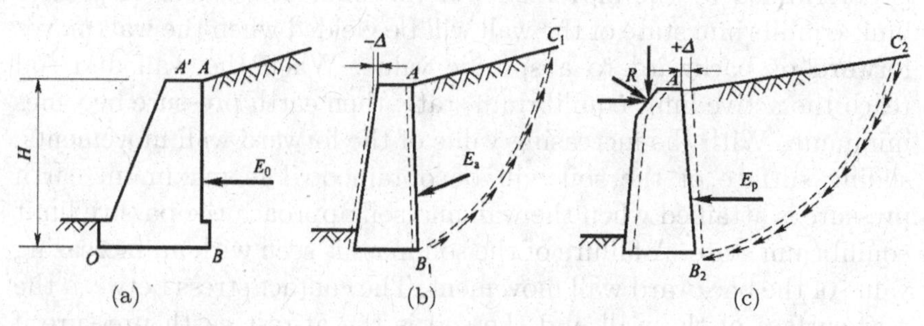


Fig. 8.2. Three types of earth pressure on the retaining wall. (a) At-rest, (b) active, and (c) passive.

as the back surface of the wall. The side opposite to the back surface is the front surface of the wall. The point $O$ is the toe of the wall, and the point $B$ is the heel of the wall. The back and front surfaces of the wall can be vertical or sloped. According to the movement condition of the retaining wall, the earth pressure on the wall can be divided into three basic types: at-rest earth pressure, active earth pressure, and passive earth pressure.

(1) **At-rest earth pressure:** If the retaining wall is static (the wall does not move either to the right or to the left of its initial position), the soil mass behind the retaining wall will be in a

state of static equilibrium. In that case, the earth pressure is referred to as the at-rest earth pressure $E_0$.

(2) **Active earth pressure:** If the frictionless wall rotates about its bottom to the position $AB_1$ (Fig. 8.2(b)), then the triangular soil mass $AB_1C_1$ adjacent to the wall will reach a critical state of equilibrium and will slide along the plane $B_1C_1$. At this time, the horizontal effective stress will be referred to as active earth pressure $E_a$.

(3) **Passive earth pressure:** If the frictionless wall rotates about its bottom to the position $AB_2$ (Fig. 8.2(c)), then the triangular soil mass $AB_2C_2$ will reach a critical state of equilibrium and will slide upward along the plane $B_2C_2$. The horizontal effective stress at this time will be the so-called passive pressure $E_{\mathrm{p}}$.

The type of earth pressure can be determined by the direction of movements of the retaining wall, and the value of the earth pressure is determined by the movements of the wall. The active or passive limit equilibrium state of the wall will be yielded when the wall moves forward or backward to a specific value. When the wall and soil reach the active limit equilibrium state, then earth pressure becomes minimum. With the increasing value of the forward wall movements, sliding surface of the soil will be obtained. The maximum earth pressure is attained when the wall and soil approach the passive limit equilibrium state. A failure of the soil mass is seen with an increasing value of the backward wall movement. The contact stress between the rear surface of the wall and the soil is the at-rest earth pressure if the wall has no movements. A comparison of the three types of earth pressures is shown in Fig. 8.3. If the active limit equilibrium state is not attained, then the actual earth pressure on the rear surface of the wall is greater than the active earth pressure $E_a$. The actual earth pressure on the rear surface of the wall is lower than the passive earth pressure $E_p$, significantly, if the passive limit equilibrium state is not attained. The at-rest earth pressure is between $E_a$ and $E_p$, i.e. $E_a < E_0 < E_p$. Sometimes in practical engineering, the conservative design of the retaining wall section (large size) is proposed for safety consideration. The wall cannot be moved and deformed; so, the earth pressure on the rear surface of the wall is the at-rest earth pressure, i.e. $E_0 > E_a$. In this case, the higher the load on the wall, more will be the risk for accident (not safe). The magnitude and distribution of

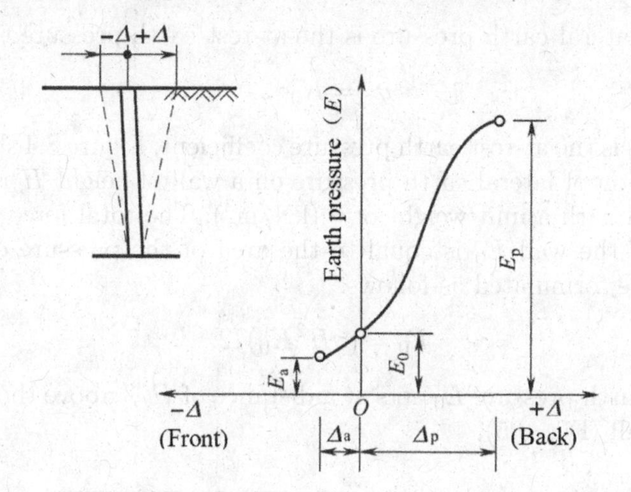


Fig. 8.3. Relationship between wall movements and earth pressure.

earth pressure have several influencing factors, e.g. the displacement conditions of the wall (including the direction and magnitude), the shape of the wall, friction condition of the rear surface of the wall, the types of the soil (e.g. sand or clay), the mechanical characteristics of the soil (e.g. shear strength), filling construction (compaction, water content, and drain conditions), and the stiffness of the wall. These factors can also influence the failure mechanism of the soil, earth pressure distribution, and the loading position.

## 8.3 Calculation of the At-Rest Earth Pressure

The retaining wall which rests on the bedrock or hard layer with a low height and high stiffness cannot be moved or rotated, and the earth pressure could be assumed as the at-rest earth pressure. The at-rest earth pressure is usually calculated with simple assumptions, that is, the rear surface of the wall is vertical and smooth, and the top surface of the soil behind the wall is flat.

### 8.3.1 *Calculation formula*

The vertical stress in the depth $z$ under the top of the wall as follows:

$$\sigma_z = \gamma z. \tag{8.1}$$

The lateral earth pressure is the at-rest earth pressure

$$\sigma_z = K_0 \gamma z, \tag{8.2}$$

where $K_0$ is the at-rest earth pressure coefficient. Figure 8.4 shows the distribution of lateral earth pressure on a wall of height $H$ retaining a dry soil with a unit weight of $\gamma$ (kN/m$^3$). The total force per unit length of the wall $E_0$ is equal to the area of the pressure diagram, $E_0$ can be formulated as follows:

$$E_0 = (\gamma H^2 K_0)/2. \tag{8.3}$$

At-rest earth pressure $E_0$ acts at a distance of $H/3$ above the bottom of the wall (Fig. 8.5).

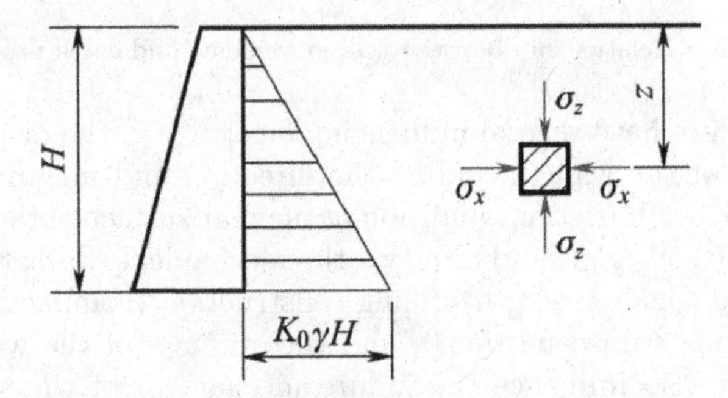


Fig. 8.4. Distribution of at-rest earth pressure.

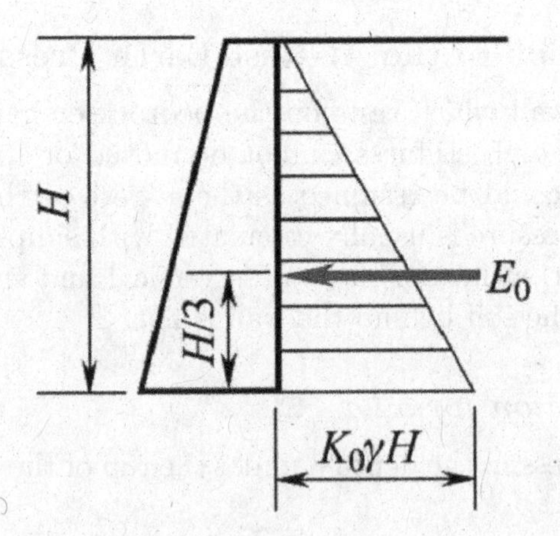


Fig. 8.5. The act point of $E_0$.

### 8.3.2 *The at-rest earth pressure coefficient* $K_0$

The at-rest earth pressure coefficient $K_0$ is the key factor of at-rest earth pressure calculation. It can be calculated via the following methods:

**(1) Elasticity method:**

$$K_0 = v/1 - v, \tag{8.4}$$

where $\nu$ (0.20–0.45) is the Poisson's ratio, $K_0 = 0.25$–$0.82$.

**(2) Jaky's method (normally consolidated gravel soil):**

An empirical relationship suggested by Jaky (1944) can estimate the at-rest earth pressure of normally consolidated (NC) gravel soils (cohesion $c = 0$), i.e.

$$K_0 = 1 - \sin\varphi', \tag{8.5}$$

where $\varphi'$ is the effective friction angle.

For fine-grained and normally consolidated soils, Massarsch (1979) suggested the following equation for $K_0$:

$$K_0 = 0.44 + 0.42 \times \left[\frac{PI(\%)}{100}\right]. \tag{8.6}$$

**(3) Overconsolidated clay:**

For overconsolidated (OC) clays, the value of $K_0$ is increased with the increasing value of overconsolidation ratio (OCR), and the at-rest earth pressure coefficient $K_0$ can be approximated as

$$K_{0(\text{OC})} = K_{0(\text{NC})}(\text{OCR})^m, \tag{8.7}$$

where $m = 0.4$–$0.5$ (Fig. 8.6).

## 8.4 Rankine's Earth Pressure Theory

Rankine (1820–1872), British scientist, published the paper "Stability of loose soil" in 1857, which was his original research work on earth pressure theory of retaining walls. He proposed the limit equilibrium state of the soil behind the wall via the half-space stress state assumption. Rankine's theory has several limitations (e.g. can be applied only for loose and cohesionless soil, $c = 0$) in

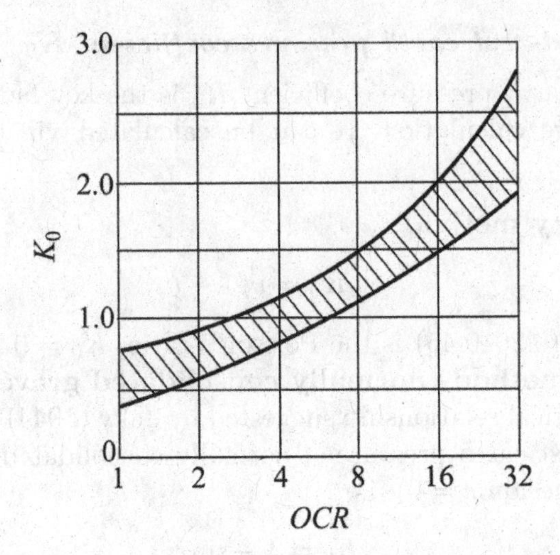


Fig. 8.6. Relationship between $K_0$ and OCR.

the evaluation of earth pressure for retaining walls without Mohr theory (Mohr, 1835–1918, German, 1900). Coulomb's earth pressure theory was established before Rankine's. Coulomb (1736–1806), a Frenchman, he proposed this earth pressure theory for retaining walls in 1773. The assumptions of Coulomb's earth pressure theory are more complicated than those of Rankine's theory, so we will discuss Rankine's theory first. At present, the Mohr limit equilibrium condition has been taken into account in Rankine's theory, so the current form of Rankine's theory is different from the original form.

### 8.4.1 *The assumptions and fundamental principles*

Assumptions are also equivalent to the application conditions. The application conditions of the Rankine earth pressure theory are as follows:

(1) The rear surface of the retaining wall is vertical and smooth.
(2) The surface of the backfilling behind the retaining wall is flat.

Each vertical surface in the semi-infinite soil is symmetrical. The shear stress on both the vertical and horizontal sections is equal to zero, so the normal stress on the corresponding sections are the principal stresses, as shown in Fig. 8.7(a). If the soil and the wall

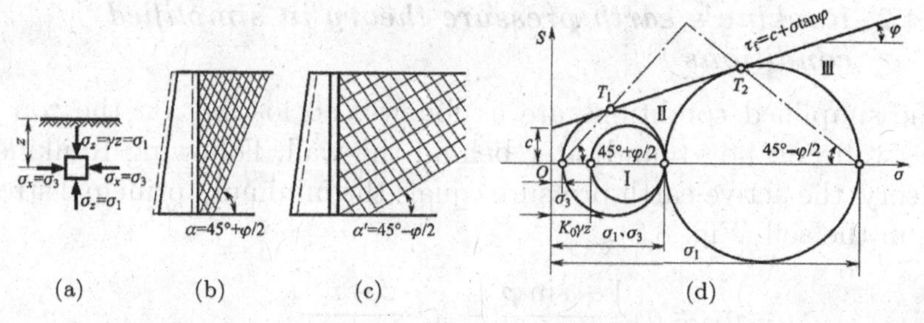


Fig. 8.7. The limiting equilibrium state of a semi-infinite soil mass. (a) Stress state, (b) active state, (c) passive state, and (d) Mohr circle.

are not moved, the contact stress on the rear surface of the wall is the at-rest earth pressure. If the shear stress on the interface of the wall and soil is zero (the rear surface of the wall is smooth), then the deformation of the soil pushes the wall to move forward. The limit equilibrium state of the retaining wall system will be attained after a certain movement of the wall (shown in Fig. 8.7(b)), and this condition is termed Rankine's active limit equilibrium state. The maximum principal stress $\sigma_1 = \sigma_z = \gamma z$ is constant, and the minimum principal stress $\sigma_3 = \sigma_x$ decreases gradually due to the relaxation state of the interface of the wall and soil. The Mohr circle moves to the-left with a decrease in the value of $\sigma_3$ until the active limit equilibrium state is reached, i.e. the Mohr circle touches the shear strength envelope. The tangent point of the Mohr circle and the shear strength envelope is the active limit equilibrium state, and the Mohr circle is named the limit Mohr's circle. The relationships between the Mohr circle and the shear strength envelope are shown in Fig. 8.7(d).

If the wall pushes the soil, the rear surface of the wall is in the state of compaction. The limit equilibrium state of the retaining wall system will also be achieved after a certain movement of the wall (shown in Fig. 8.7(c)), and this condition is named Rankine's passive limit equilibrium state. In this condition, the lateral stress $\sigma_x$ turns to the maximum principal stress $\sigma_1$, and it will increase to a high magnitude until the passive limit equilibrium state is achieved. The Mohr circle keeps moving to the right and achieves the passive limit equilibrium state. The maximum principal stress $\sigma_1$ is equal to the lateral stress $\sigma_x$ and is also named the Rankine passive earth pressure.

### 8.4.2 *Rankine's earth pressure theory in simplified conditions*

The simplified conditions are as follows: no loading on the top of the wall and no groundwater behind the wall. Following Rankine's theory, the active earth pressure equals the minimum principal stress $\sigma_3$ in the soil (Fig. 8.8):

$$\begin{aligned}\sigma_3 = \sigma_a &= \sigma_1 \frac{1-\sin\varphi}{1+\sin\varphi} - 2c\frac{\cos\varphi}{1+\sin\varphi}\\ &= \gamma z \tan^2\left(45^\circ - \frac{\varphi}{2}\right) - 2c\tan\left(45^\circ - \frac{\varphi}{2}\right). \end{aligned} \tag{8.8}$$

Equation (8.8) can be divided into two parts: the first part, $\gamma z \tan^2(45^\circ - \varphi/2)$, is a triangle *oec*, and the second part, $2c\tan^2(45^\circ - \varphi/2)$, is a quadrangle *oedb*. If the area of the triangle *oec* is subtracted from the area of the quadrangle *oedb*, the area of the triangle *abc* will be obtained. The total force of active earth pressure acting on the rear surface of the wall is the area of the triangle *abc* (sum of the earth pressure distribution). The location $(H - z_0)$ of the acting point of the total force $E_0$ can be calculated by Eq. (8.8) with $z = z_0$.

$$\sigma_3 = \gamma z_0 \tan^2\left(45^\circ - \frac{\varphi}{2}\right) - 2c\tan\left(45^\circ - \frac{\varphi}{2}\right) = 0, \tag{8.9}$$

$$z_0 = \frac{2c}{\gamma \tan\left(45^\circ - \frac{\varphi}{2}\right)}. \tag{8.10}$$

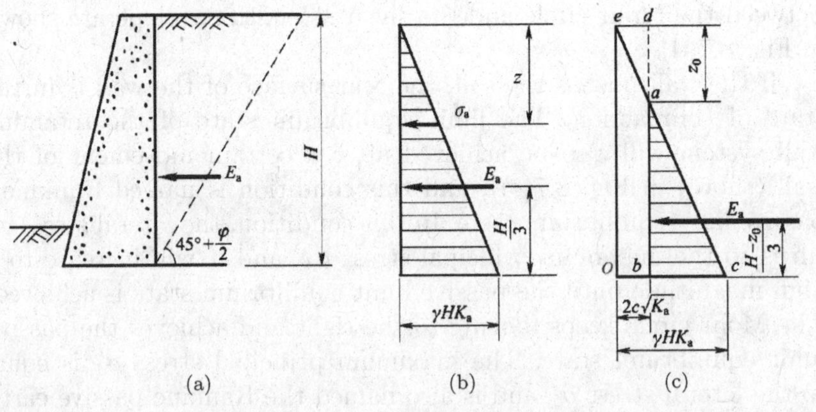


Fig. 8.8. Distribution of Rankine's active earth pressure. (a) Retaining wall, (b) cohesionless soil, and (c) cohesive soil.

Therefore, the total force $E_0$ (equal to the area of triangle *abc*) is

$$E_a = \frac{1}{2}bc \times ab = \frac{1}{2}\left[\gamma H \tan^2\left(45^\circ - \frac{\varphi}{2}\right) - 2c\tan\left(45^\circ - \frac{\varphi}{2}\right)\right]$$
$$\times(H - z_0)$$
$$= \frac{1}{2}\gamma H^2 \tan^2\left(45^\circ - \frac{\varphi}{2}\right) - 2cH\tan\left(45^\circ - \frac{\varphi}{2}\right) + \frac{2c^2}{\gamma}. \quad (8.11)$$

From Eq. (8.11), it can be seen that the unit of $E_a$ is kN/m, i.e. $E_a$ is the force acting on the unit length of the rear surface of the wall. Because the rear surface of the wall is smooth, $E_a$ can only act in the horizontal direction. The action point of $E_a$ is at the centroid of the triangle *abc*, i.e. one-third of *ab*, $(H - z_0)/3$, away from the bottom of the wall. The active earth pressure coefficient $K_a$ can be formulated as

$$K_a = \frac{\sigma_3}{\sigma_1} = \frac{\sigma_3}{\gamma z}. \quad (8.12)$$

If the cohesion $c$ equals zero in Eq. (8.8) (cohesionless soil), then the active earth pressure coefficient $K_a$ can be written as

$$K_a = \tan^2\left(45^\circ - \frac{\varphi}{2}\right). \quad (8.13)$$

The minimum principal stress $\sigma_3$ will be equal to $\gamma z$ under the passive limit equilibrium state, and the maximum principal stress $\sigma_1$ can be obtained as follows:

$$\sigma_1 = \sigma_p = \sigma_3\frac{1+\sin\varphi}{1-\sin\varphi} + 2c\frac{\cos\varphi}{1-\sin\varphi}$$
$$= \gamma z \tan^2\left(45^\circ + \frac{\varphi}{2}\right) + 2c\tan\left(45^\circ + \frac{\varphi}{2}\right). \quad (8.14)$$

Equation (8.14) can also be divided into two parts: the first part, $\gamma z\tan^2(45^\circ + \varphi/2)$, is a triangle, and the second part, $2c\tan^2(45^\circ + \varphi/2)$, is a quadrangle. Adding the two parts together, we obtain a trapezoid *abcd*. The passive earth pressure acting on the rear surface of the wall is the area of trapezoid *abcd* (sum of the earth pressure

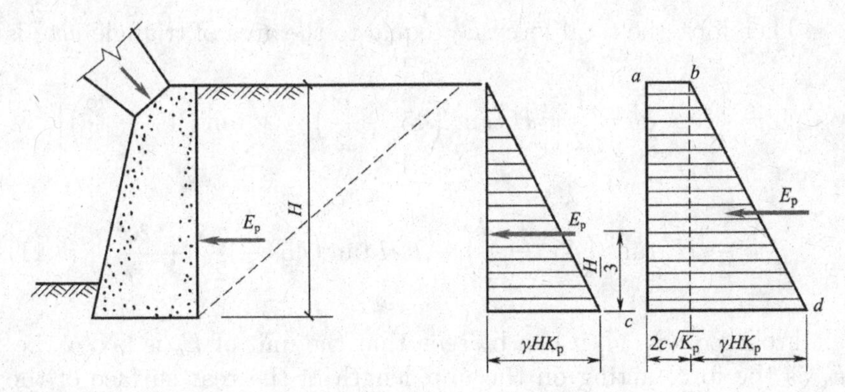


Fig. 8.9. Distribution of Rankine's passive earth pressure. (a) Retaining wall, (b) cohesionless soil, and (c) cohesive soil.

distribution) (Fig. 8.9).

$$E_p = \frac{1}{2}H(ab + cd) = \frac{1}{2}H\left[\gamma H\tan^2\left(45^\circ + \frac{\varphi}{2}\right) + 4c\tan\left(45^\circ + \frac{\varphi}{2}\right)\right]$$

$$= \frac{1}{2}\gamma H^2\tan^2\left(45^\circ + \frac{\varphi}{2}\right) + 2cH\tan\left(45^\circ + \frac{\varphi}{2}\right). \tag{8.15}$$

If the cohesion $c$ is equal to zero in Eq. (8.14) (cohesionless soil), then the passive earth pressure coefficient $K_P$ can be written as

$$K_P = \tan^2\left(45^\circ + \frac{\varphi}{2}\right). \tag{8.16}$$

**Example 8.1.** Calculate the total active thrust on a vertical wall 5 m high retaining a sand of unit weight $17\,\text{kN/m}^3$; for which $\varphi' = 35^\circ$; the surface of the sand is horizontal and the water table is below the bottom of the wall. Determine the thrust on the wall if the water table rises to a level 2 m below the surface of the sand. The saturated unit weight of the sand is $20\,\text{kN/m}^3$ (Fig. 8.10).

**Solution.**

(a)

$$K_a = \frac{1 - \sin 35^\circ}{1 + \sin 35^\circ} = 0.27,$$

$$E_a = \frac{1}{2}K_a\gamma H^2 = 57.4\,\text{kN/m}.$$

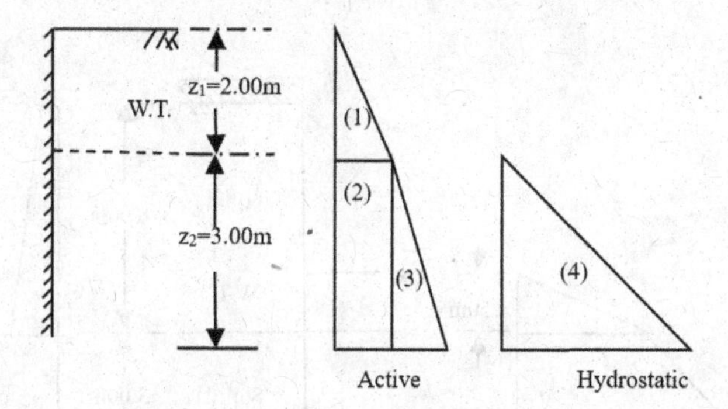


Fig. 8.10. Example 8.1.

(b) The pressure distribution on the wall is now as shown in Fig. 8.10, including hydrostatic pressure on the lower 3 m of the wall.

Above the water table, the active earth pressure is

$$P_{a1} = K_a \gamma z_1 = 0.27 \times 17 \times 2 = 9.18\,\text{kPa}.$$

Below the water table, the active earth pressure should be calculated in terms of the effective weight of the soil, so the active earth pressure at the bottom of the wall is obtained as follows:

$$P_{a2} = P_{a1} + Ka\gamma' z_2 = 9.18 + 0.27 \times (20 - 9.8) \times 3 = 17.44\,\text{kPa}.$$

The water pressure is thus given as

$$P_w = \gamma_w z_2 = 9.8 \times 3 = 29.4\,\text{kPa}.$$

Therefore, the total thrust is (Fig. 8.10) given as

$$E_a = \frac{1}{2} \times 9.18 \times 2 + \frac{1}{2} \times (9.18 + 17.44) \times 3 + \frac{1}{2} \times 29.4 \times 3 = 93.21\,\text{kN/m}.$$

**Example 8.2.** The soil conditions adjacent to a sheet pile wall are given in Fig. 8.11, and a surcharge pressure of $50\,\text{kN/m}^2$ is carried on the surface behind the wall. For soil 1, i.e. sand above the water table, $c' = 0$, $\varphi' = 38°$, and $\gamma = 18\,\text{kN/m}^3$. For soil 2, i.e. saturated clay, $c' = 10\,\text{kN/m}^2$, $\varphi' = 28°$, and $\gamma_{\text{sat}} = 20\,\text{kN/m}^3$. Plot the distributions of active pressure behind the wall and passive pressure in front of the wall.

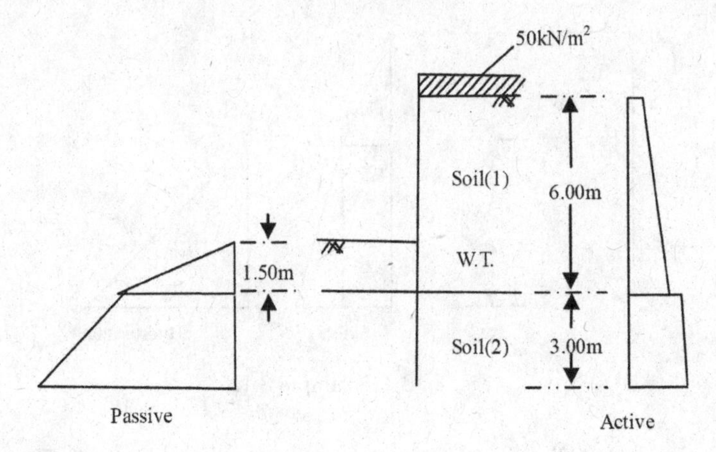


Fig. 8.11. Example 8.2.

**Solution.**

For soil 1:

$$K_a = \frac{1 - \sin 38^\circ}{1 + \sin 38^\circ} = 0.24, \quad K_P = \frac{1}{0.24} = 4.17.$$

The pressures in soil 1 are calculated using $K_a = 0.24$, $K_p = 4.17$, and $\gamma = 18\,\text{kN/m}^3$. Soil 1 is then considered as a surcharge of $(18 \times 6)\,\text{kN/m}^2$ on soil 2, in addition to the surface surcharge. The pressures in soil 2 are calculated using $K_a = 0.36$, $K_p = 2.78$, and $\gamma' = (20 - 9.8) = 10.2\,\text{kN/m}^3$ (see Table 8.1). The active and passive pressure distributions are shown in Fig. 8.11. In addition, there is equal hydrostatic pressure on each side of the wall below the water table.

## 8.5 Coulomb's Earth Pressure Theory

Coulomb's theory of earth pressure is proposed before Rankine's earth pressure theory, and it is established by Coulomb (1773) using the static equilibrium condition of the sliding wedge soil mass behind the retaining wall under the limit equilibrium state. The polygon force equilibrium principle of theoretical mechanics is used to deduce Coulomb's earth pressure theory, so this theory is simple and widely used. However, the application conditions of Coulomb's theory are more complex than Rankine's theory.

Table 8.1. Calculation procedure for Example 8.2.

| Soil | Depth (m) | Pressure (kN/m$^2$) | |
|---|---|---|---|
| *Active pressure* | | | |
| 1 | 0 | $0.24 \times 50$ | $= 12.0$ |
| 1 | 6 | $(0.24 \times 50) + (0.24 \times 18 \times 6)$ | $= 37.9$ |
| 2 | 6 | $0.36(50 + (18 \times 6)) - (2 \times 10 \times \sqrt{0.36})$ | $= 44.9$ |
| 2 | 9 | $0.36(50 + (18 \times 6)) - (2 \times 10 \times \sqrt{0.36})$ $+ (0.36 \times 10.2 \times 3)$ | $= 55.9$ |
| *Passive pressure* | | | |
| 1 | 0 | 0 | |
| 1 | 1.5 | $4.17 \times 18 \times 1.5$ | $= 112.6$ |
| 2 | 1.5 | $(2.78 \times 18 \times 1.5) + (2 \times 10 \times \sqrt{2.78})$ | $= 108.4$ |
| 2 | 4.5 | $(2.78 \times 18 \times 1.5) + (2 \times 10 \times \sqrt{2.78})$ $+ (2.78 \times 10.2 \times 3)$ | $= 193.5$ |

### 8.5.1 *Assumptions and application conditions*

#### 8.5.1.1 *Assumptions*

(1) The filling behind the wall is granular soil, e.g. sand (cohesion $c = 0$).
(2) An oblique plane of slip surface will be yielded through the back surface and heel of the wall when the rear surface of the wall moves forward or backward and reaches the limit equilibrium state. The wedge $ABM$ is a rigid block and slides along the slip surface $AM$.
(3) The wedge $ABM$ is in the limit equilibrium state, and it can be analyzed via the rigid equilibrium method of theoretical mechanics, i.e. the polygon force equilibrium principle.

#### 8.5.1.2 *Application conditions (shown in Fig. 8.12)*

(1) The rear surface of the wall is inclined, and the dip angle is $\alpha$.
(2) The rear surface of the wall is rough, and the friction angle of the interface between the wall and the soil is $\delta$.
(3) The surface of the filling behind the wall is inclined, and the slope angle is $\beta$.

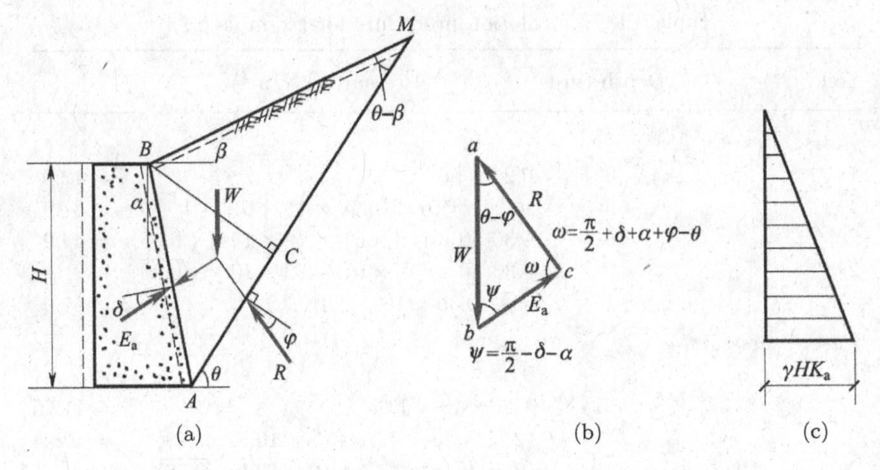


Fig. 8.12. Scheme for calculating Coulomb's active earth pressure.

## 8.5.2 *Derivation of the formula for Coulomb's active earth pressure calculation*

### 8.5.2.1 *Formula derivation*

As shown in Fig. 8.12(a), the rigid wedge soil $\Delta ABM$ is in a limit equilibrium state. The magnitude of reaction force $R$ on the surface $AM$ is uncertain, but the direction is fixed (the angle between the normal of the surface $AM$ and reaction force $R$ is $\varphi$). The reaction force $E_a$ on the rear surface of the wall $AB$ is uncertain, but the direction is also fixed (the angle between the normal of the surface $AB$ and reaction force $E_a$ is $\delta$). The direction of the friction force on $AM$ is opposite for the motion of $ABM$. Using the polygon force equilibrium principle, the geometrical relationship of those forces is shown in Fig. 8.12(b). The gravity of the wedge $ABMW = \gamma \cdot AM \cdot BC/2$, and $AM$ and $BC$ can be formulated using the trigonometric relation as follows:

$$AM = AB\frac{\sin[90^\circ - (\alpha - \beta)]}{\sin(\theta - \beta)} = \frac{\cos(\alpha - \beta)}{\sin(\theta - \beta)},$$

$$AB = \frac{H}{\cos\alpha},$$

$$BC = AB\cos\angle ABC = AB\cos(\theta - \alpha) = H \bullet \frac{\cos(\theta - \alpha)}{\cos\alpha}.$$

The gravity $W$ of the wedge $ABM$ can be written as

$$W = \frac{1}{2}\gamma H^2 \frac{\cos(\alpha - \beta)\cos(\theta - \alpha)}{\cos^2\alpha \sin(\theta - \beta)}.$$

The reaction force $E_a$ in $\Delta abc$ (see Fig. 8.12(b)) can be expressed as

$$\frac{E_a}{\sin(\theta - \varphi)} = \frac{W}{\sin\omega},$$

$$E_a = W\frac{\sin(\theta - \varphi)}{\sin\omega}$$

$$= \frac{1}{2}\gamma H^2 \frac{\cos(\alpha - \beta)\cos(\theta - \alpha)\sin(\theta - \varphi)}{\cos^2\alpha \sin(\theta - \beta)\sin(\delta + \alpha + \varphi - \theta)}.$$

The reaction force $E_a$ can also be written by $K_a$ as follows:

$$E_a = \frac{1}{2}K_a\gamma H^2, \tag{8.17}$$

$$K_a = \frac{\cos^2(\varphi - \alpha)}{\cos^2\alpha\cos(\alpha + \delta)\left[1 + \sqrt{\frac{\sin(\varphi+\delta)\sin(\varphi-\beta)}{\cos(\alpha+\delta)\cos(\alpha-\beta)}}\right]^2}. \tag{8.18}$$

The distribution of $E_a$ along the wall height $H$ is shown in Fig. 8.12(c). The acting point of $E_a$ is located at one-third of the wall height $H$, i.e. $H/3$ away from the bottom of the wall, and it also can be approximated as one-third of the rear surface of the wall away from the bottom. $K_a$ can be obtained by the interpolation method, as shown in Table 8.2.

#### 8.5.2.2 *Notes for the application of Eqs. (8.17) and (8.18)*

(1) There are two conditions for the incline in the retaining wall (see Fig. 8.13).

By replacing the value of the angle $\alpha$ with the symbol $\pm$ into Eqs. (8.17) and (8.18), we can calculate $E_a$ and $K_a$. However, the value of the angle $\alpha$ can't be too large, and the reasonable value of the angle $\alpha$ should not exceed $\pm 20°$.

(2) The value of the friction angle $\delta$ between the rear surface of the wall and the soil is shown in Table 8.3.

(3) If $K_a = \tan^2(45° - \varphi/2)$ when $\alpha = \beta = \delta = 0$, then the expression of $K_a$ is the same as that of Rankine's theory.

Table 8.2. Active earth pressure coefficient $K_a$.

| $\alpha$ | $\beta$ | $\varphi$ 15° | 20° | 25° | 30° | 35° | 40° | 45° | 50° |
|---|---|---|---|---|---|---|---|---|---|
| | | | | | $\delta = 0°$ | | | | |
| | 0° | 0.589 | 0.490 | 0.406 | 0.333 | 0.271 | 0.217 | 0.172 | 0.132 |
| | 5° | 0.635 | 0.524 | 0.431 | 0.352 | 0.284 | 0.227 | 0.178 | 0.137 |
| | 10° | 0.704 | 0.569 | 0.462 | 0.374 | 0.300 | 0.238 | 0.186 | 0.142 |
| | 15° | 0.933 | 0.639 | 0.505 | 0.402 | 0.319 | 0.251 | 0.194 | 0.147 |
| | 20° | | 0.883 | 0.573 | 0.441 | 0.344 | 0.267 | 0.204 | 0.154 |
| 0° | 25° | | | 0.821 | 0.505 | 0.379 | 0.288 | 0.217 | 0.162 |
| | 30° | | | | 0.750 | 0.436 | 0.318 | 0.235 | 0.172 |
| | 35° | | | | | 0.671 | 0.369 | 0.260 | 0.186 |
| | 40° | | | | | | 0.587 | 0.303 | 0.206 |
| | 45° | | | | | | | 0.500 | 0.242 |
| | 50° | | | | | | | | 0.143 |
| | 0° | 0.652 | 0.560 | 0.478 | 0.407 | 0.343 | 0.288 | 0.238 | 0.194 |
| | 5° | 0.705 | 0.601 | 0.510 | 0.431 | 0.362 | 0.302 | 0.249 | 0.202 |
| | 10° | 0.784 | 0.655 | 0.550 | 0.461 | 0.384 | 0.318 | 0.261 | 0.211 |
| | 15° | 1.039 | 0.737 | 0.603 | 0.498 | 0.411 | 0.337 | 0.274 | 0.221 |
| | 20° | | 1.015 | 0.685 | 0.548 | 0.444 | 0.360 | 0.291 | 0.231 |
| 10° | 25° | | | 0.977 | 0.628 | 0.491 | 0.391 | 0.311 | 0.245 |
| | 30° | | | | 0.925 | 0.566 | 0.433 | 0.337 | 0.262 |
| | 35° | | | | | 0.860 | 0.502 | 0.374 | 0.284 |
| | 40° | | | | | | 0.785 | 0.437 | 0.316 |
| | 45° | | | | | | | 0.703 | 0.371 |
| | 50° | | | | | | | | 0.614 |
| | 0° | 0.736 | 0.648 | 0.569 | 0.498 | 0.434 | 0.375 | 0.322 | 0.274 |
| | 5° | 0.801 | 0.700 | 0.611 | 0.532 | 0.461 | 0.397 | 0.340 | 0.288 |
| | 10° | 0.896 | 0.768 | 0.663 | 0.572 | 0.492 | 0.421 | 0.358 | 0.302 |
| | 15° | 1.196 | 0.868 | 0.730 | 0.621 | 0.529 | 0.450 | 0.380 | 0.318 |
| | 20° | | 1.205 | 0.834 | 0.688 | 0.576 | 0.484 | 0.405 | 0.337 |
| 20° | 25° | | | 1.196 | 0.791 | 0.639 | 0.527 | 0.435 | 0.358 |
| | 30° | | | | 1.169 | 0.740 | 0.586 | 0.474 | 0.385 |
| | 35° | | | | | 1.124 | 0.683 | 0.529 | 0.420 |
| | 40° | | | | | | 1.064 | 0.620 | 0.469 |
| | 45° | | | | | | | 0.990 | 0.552 |
| | 50° | | | | | | | | 0.904 |

(*Continued*)

Table 8.2. (*Continued*)

| $\alpha$ | $\beta$ | $\varphi$ | | | | | | | |
|---|---|---|---|---|---|---|---|---|---|
| | | 15° | 20° | 25° | 30° | 35° | 40° | 45° | 50° |
| | | | | | $\delta = 0°$ | | | | |
| | 0° | 0.540 | 0.433 | 0.344 | 0.270 | 0.209 | 0.158 | 0.117 | 0.083 |
| | 5° | 0.581 | 0.461 | 0.364 | 0.284 | 0.218 | 0.164 | 0.120 | 0.085 |
| | 10° | 0.644 | 0.500 | 0.389 | 0.301 | 0.229 | 0.171 | 0.125 | 0.088 |
| | 15° | 0.860 | 0.562 | 0.425 | 0.322 | 0.243 | 0.180 | 0.130 | 0.090 |
| | 20° | | 0.785 | 0.482 | 0.353 | 0.261 | 0.190 | 0.136 | 0.094 |
| −10° | 25° | | | 0.703 | 0.405 | 0.287 | 0.205 | 0.144 | 0.098 |
| | 30° | | | | 0.614 | 0.331 | 0.226 | 0.155 | 0.104 |
| | 35° | | | | | 0.523 | 0.263 | 0.171 | 0.111 |
| | 40° | | | | | | 0.433 | 0.200 | 0.123 |
| | 45° | | | | | | | 0.344 | 0.145 |
| | 50° | | | | | | | | 0.262 |
| | 0° | 0.497 | 0.380 | 0.287 | 0.212 | 0.153 | 0.106 | 0.070 | 0.043 |
| | 5° | 0.535 | 0.405 | 0.302 | 0.222 | 0.159 | 0.110 | 0.072 | 0.044 |
| | 10° | 0.595 | 0.439 | 0.323 | 0.234 | 0.166 | 0.114 | 0.074 | 0.045 |
| | 15° | 0.809 | 0.494 | 0.352 | 0.250 | 0.175 | 0.119 | 0.076 | 0.046 |
| | 20° | | 0.707 | 0.401 | 0.274 | 0.188 | 0.125 | 0.080 | 0.047 |
| −20° | 25° | | | 0.603 | 0.316 | 0.206 | 0.134 | 0.084 | 0.049 |
| | 30° | | | | 0.498 | 0.239 | 0.147 | 0.090 | 0.051 |
| | 35° | | | | | 0.396 | 0.172 | 0.099 | 0.055 |
| | 40° | | | | | | 0.301 | 0.116 | 0.060 |
| | 45° | | | | | | | 0.215 | 0.071 |
| | 50° | | | | | | | | 0.141 |
| | | | | | $\delta = 5°$ | | | | |
| | 0° | 0.556 | 0.465 | 0.387 | 0.319 | 0.260 | 0.210 | 0.166 | 0.129 |
| | 5° | 0.605 | 0.500 | 0.412 | 0.337 | 0.274 | 0.219 | 0.173 | 0.133 |
| | 10° | 0.680 | 0.547 | 0.444 | 0.360 | 0.289 | 0.230 | 0.180 | 0.138 |
| | 15° | 0.937 | 0.620 | 0.488 | 0.388 | 0.308 | 0.243 | 0.189 | 0.144 |
| | 20° | | 0.886 | 0.558 | 0.428 | 0.333 | 0.259 | 0.199 | 0.150 |
| 0° | 25° | | | 0.825 | 0.493 | 0.369 | 0.280 | 0.212 | 0.158 |
| | 30° | | | | 0.753 | 0.428 | 0.311 | 0.229 | 0.168 |
| | 35° | | | | | 0.674 | 0.363 | 0.255 | 0.182 |
| | 40° | | | | | | 0.589 | 0.299 | 0.202 |
| | 45° | | | | | | | 0.502 | 0.388 |
| | 50° | | | | | | | | 0.415 |

(*Continued*)

Table 8.2. (*Continued*)

| $\alpha$ | $\beta$ | $\varphi$ 15° | 20° | 25° | 30° | 35° | 40° | 45° | 50° |
|---|---|---|---|---|---|---|---|---|---|
| | | | | | $\delta = 5°$ | | | | |
| | 0° | 0.622 | 0.536 | 0.460 | 0.393 | 0.333 | 0.280 | 0.233 | 0.191 |
| | 5° | 0.680 | 0.579 | 0.493 | 0.418 | 0.352 | 0.294 | 0.243 | 0.199 |
| | 10° | 0.767 | 0.636 | 0.534 | 0.448 | 0.374 | 0.311 | 0.255 | 0.207 |
| | 15° | 1.060 | 0.725 | 0.589 | 0.486 | 0.401 | 0.330 | 0.269 | 0.217 |
| | 20° | | 1.035 | 0.676 | 0.538 | 0.436 | 0.354 | 0.286 | 0.228 |
| 10° | 25° | | | 0.996 | 0.622 | 0.484 | 0.385 | 0.306 | 0.242 |
| | 30° | | | | 0.943 | 0.563 | 0.428 | 0.333 | 0.259 |
| | 35° | | | | | 0.877 | 0.500 | 0.371 | 0.281 |
| | 40° | | | | | | 0.801 | 0.436 | 0.314 |
| | 45° | | | | | | | 0.716 | 0.371 |
| | 50° | | | | | | | | 0.626 |
| | 0° | 0.709 | 0.627 | 0.553 | 0.485 | 0.424 | 0.368 | 0.318 | 0.271 |
| | 5° | 0.781 | 0.682 | 0.597 | 0.520 | 0.452 | 0.391 | 0.335 | 0.285 |
| | 10° | 0.887 | 0.755 | 0.650 | 0.562 | 0.484 | 0.416 | 0.355 | 0.300 |
| | 15° | 1.240 | 0.866 | 0.723 | 0.614 | 0.523 | 0.445 | 0.376 | 0.316 |
| | 20° | | 1.250 | 0.835 | 0.684 | 0.571 | 0.480 | 0.402 | 0.335 |
| 20° | 25° | | | 1.240 | 0.794 | 0.639 | 0.525 | 0.434 | 0.357 |
| | 30° | | | | 1.212 | 0.746 | 0.587 | 0.474 | 0.385 |
| | 35° | | | | | 1.166 | 0.689 | 0.532 | 0.421 |
| | 40° | | | | | | 1.103 | 0.627 | 0.472 |
| | 45° | | | | | | | 1.026 | 0.559 |
| | 50° | | | | | | | | 0.937 |
| | 0° | 0.503 | 0.406 | 0.324 | 0.256 | 0.199 | 0.151 | 0.112 | 0.080 |
| | 5° | 0.546 | 0.434 | 0.344 | 0.269 | 0.208 | 0.157 | 0.116 | 0.082 |
| | 10° | 0.612 | 0.474 | 0.369 | 0.286 | 0.219 | 0.164 | 0.120 | 0.085 |
| | 15° | 0.850 | 0.537 | 0.405 | 0.308 | 0.232 | 0.172 | 0.125 | 0.087 |
| | 20° | | 0.776 | 0.463 | 0.339 | 0.250 | 0.183 | 0.131 | 0.091 |
| −10° | 25° | | | 0.695 | 0.390 | 0.276 | 0.197 | 0.139 | 0.095 |
| | 30° | | | | 0.607 | 0.321 | 0.218 | 0.149 | 0.100 |
| | 35° | | | | | 0.518 | 0.255 | 0.166 | 0.108 |
| | 40° | | | | | | 0.428 | 0.195 | 0.120 |
| | 45° | | | | | | | 0.341 | 0.141 |
| | 50° | | | | | | | | 0.259 |

(*Continued*)

Table 8.2. (*Continued*)

| $\alpha$ | $\beta$ | $\varphi$ 15° | 20° | 25° | 30° | 35° | 40° | 45° | 50° |
|---|---|---|---|---|---|---|---|---|---|
| | | | | | $\delta = 5°$ | | | | |
| | 0° | 0.457 | 0.352 | 0.267 | 0.199 | 0.144 | 0.101 | 0.067 | 0.041 |
| | 5° | 0.496 | 0.376 | 0.282 | 0.208 | 0.150 | 0.104 | 0.068 | 0.042 |
| | 10° | 0.557 | 0.410 | 0.302 | 0.220 | 0.157 | 0.108 | 0.070 | 0.043 |
| | 15° | 0.787 | 0.466 | 0.331 | 0.236 | 0.165 | 0.112 | 0.073 | 0.044 |
| | 20° | | 0.688 | 0.380 | 0.259 | 0.178 | 0.119 | 0.076 | 0.045 |
| −20° | 25° | | | 0.586 | 0.300 | 0.196 | 0.127 | 0.080 | 0.047 |
| | 30° | | | | 0.484 | 0.228 | 0.140 | 0.085 | 0.049 |
| | 35° | | | | | 0.386 | 0.165 | 0.094 | 0.052 |
| | 40° | | | | | | 0.293 | 0.111 | 0.058 |
| | 45° | | | | | | | 0.209 | 0.068 |
| | 50° | | | | | | | | 0.137 |
| | | | | | $\delta = 10°$ | | | | |
| | 0° | 0.533 | 0.447 | 0.373 | 0.309 | 0.253 | 0.204 | 0.163 | 0.127 |
| | 5° | 0.585 | 0.483 | 0.398 | 0.327 | 0.266 | 0.214 | 0.169 | 0.131 |
| | 10° | 0.664 | 0.531 | 0.431 | 0.350 | 0.282 | 0.225 | 0.177 | 0.136 |
| | 15° | 0.947 | 0.609 | 0.476 | 0.379 | 0.301 | 0.238 | 0.185 | 0.141 |
| | 20° | | 0.897 | 0.549 | 0.420 | 0.326 | 0.254 | 0.195 | 0.148 |
| 0° | 25° | | | 0.834 | 0.487 | 0.363 | 0.275 | 0.209 | 0.156 |
| | 30° | | | | 0.762 | 0.423 | 0.306 | 0.226 | 0.166 |
| | 35° | | | | | 0.681 | 0.359 | 0.252 | 0.180 |
| | 40° | | | | | | 0.596 | 0.297 | 0.201 |
| | 45° | | | | | | | 0.508 | 0.238 |
| | 50° | | | | | | | | 0.420 |
| | 0° | 0.603 | 0.520 | 0.448 | 0.384 | 0.326 | 0.275 | 0.230 | 0.189 |
| | 5° | 0.665 | 0.566 | 0.482 | 0.409 | 0.346 | 0.290 | 0.240 | 0.197 |
| | 10° | 0.759 | 0.626 | 0.524 | 0.440 | 0.369 | 0.307 | 0.253 | 0.206 |
| | 15° | 1.089 | 0.721 | 0.582 | 0.480 | 0.396 | 0.326 | 0.267 | 0.216 |
| | 20° | | 1.064 | 0.674 | 0.534 | 0.432 | 0.351 | 0.284 | 0.227 |
| 10° | 25° | | | 1.024 | 0.622 | 0.482 | 0.382 | 0.304 | 0.241 |
| | 30° | | | | 0.969 | 0.564 | 0.427 | 0.332 | 0.258 |
| | 35° | | | | | 0.901 | 0.503 | 0.371 | 0.281 |
| | 40° | | | | | | 0.823 | 0.438 | 0.315 |
| | 45° | | | | | | | 0.736 | 0.374 |
| | 50° | | | | | | | | 0.644 |

(*Continued*)

Table 8.2. (*Continued*)

| $\alpha$ | $\beta$ | 15° | 20° | 25° | 30° | 35° | 40° | 45° | 50° |
|---|---|---|---|---|---|---|---|---|---|
| | | | | | $\delta = 10°$ | | | | |
| | 0° | 0.695 | 0.615 | 0.543 | 0.478 | 0.419 | 0.365 | 0.316 | 0.271 |
| | 5° | 0.773 | 0.674 | 0.589 | 0.515 | 0.448 | 0.388 | 0.334 | 0.285 |
| | 10° | 0.890 | 0.752 | 0.646 | 0.558 | 0.482 | 0.414 | 0.354 | 0.300 |
| | 15° | 1.298 | 0.872 | 0.723 | 0.613 | 0.522 | 0.444 | 0.377 | 0.317 |
| | 20° | | 1.308 | 0.844 | 0.687 | 0.573 | 0.481 | 0.403 | 0.337 |
| 20° | 25° | | | 1.298 | 0.806 | 0.643 | 0.528 | 0.436 | 0.360 |
| | 30° | | | | 1.268 | 0.758 | 0.594 | 0.478 | 0.388 |
| | 35° | | | | | 1.220 | 0.702 | 0.539 | 0.426 |
| | 40° | | | | | | 1.155 | 0.640 | 0.480 |
| | 45° | | | | | | | 1.074 | 0.572 |
| | 50° | | | | | | | | 0.981 |
| | 0° | 0.477 | 0.385 | 0.309 | 0.245 | 0.191 | 0.146 | 0.109 | 0.078 |
| | 5° | 0.521 | 0.414 | 0.329 | 0.258 | 0.200 | 0.152 | 0.112 | 0.080 |
| | 10° | 0.590 | 0.455 | 0.354 | 0.275 | 0.211 | 0.159 | 0.116 | 0.082 |
| | 15° | 0.847 | 0.520 | 0.390 | 0.297 | 0.224 | 0.167 | 0.121 | 0.085 |
| | 20° | | 0.773 | 0.450 | 0.328 | 0.242 | 0.177 | 0.127 | 0.088 |
| −10° | 25° | | | 0.692 | 0.380 | 0.268 | 0.191 | 0.135 | 0.093 |
| | 30° | | | | 0.605 | 0.313 | 0.212 | 0.146 | 0.098 |
| | 35° | | | | | 0.516 | 0.249 | 0.162 | 0.106 |
| | 40° | | | | | | 0.426 | 0.191 | 0.117 |
| | 45° | | | | | | | 0.339 | 0.139 |
| | 50° | | | | | | | | 0.258 |
| | 0° | 0.427 | 0.330 | 0.252 | 0.188 | 0.137 | 0.096 | 0.064 | 0.039 |
| | 5° | 0.466 | 0.354 | 0.267 | 0.197 | 0.143 | 0.099 | 0.066 | 0.040 |
| | 10° | 0.529 | 0.388 | 0.286 | 0.209 | 0.149 | 0.103 | 0.068 | 0.041 |
| | 15° | 0.772 | 0.445 | 0.315 | 0.225 | 0.158 | 0.108 | 0.070 | 0.042 |
| | 20° | | 0.675 | 0.364 | 0.248 | 0.170 | 0.114 | 0.073 | 0.044 |
| −20° | 25° | | | 0.575 | 0.288 | 0.188 | 0.122 | 0.077 | 0.045 |
| | 30° | | | | 0.475 | 0.220 | 0.135 | 0.082 | 0.047 |
| | 35° | | | | | 0.378 | 0.159 | 0.091 | 0.051 |
| | 40° | | | | | | 0.288 | 0.108 | 0.056 |
| | 45° | | | | | | | 0.205 | 0.066 |
| | 50° | | | | | | | | 0.135 |

(*Continued*)

Table 8.2. (*Continued*)

| $\alpha$ | $\beta$ | 15° | 20° | 25° | 30° | 35° | 40° | 45° | 50° |
|---|---|---|---|---|---|---|---|---|---|
| | | | | | $\delta = 15°$ | | | | |
| | 0° | 0.518 | 0.434 | 0.363 | 0.301 | 0.248 | 0.201 | 0.160 | 0.125 |
| | 5° | 0.571 | 0.471 | 0.389 | 0.320 | 0.261 | 0.211 | 0.167 | 0.130 |
| | 10° | 0.656 | 0.522 | 0.423 | 0.343 | 0.277 | 0.222 | 0.174 | 0.135 |
| | 15° | 0.966 | 0.603 | 0.470 | 0.373 | 0.297 | 0.235 | 0.183 | 0.140 |
| | 20° | | 0.914 | 0.546 | 0.415 | 0.323 | 0.251 | 0.194 | 0.147 |
| 0° | 25° | | | 0.850 | 0.485 | 0.360 | 0.273 | 0.207 | 0.155 |
| | 30° | | | | 0.777 | 0.422 | 0.305 | 0.225 | 0.165 |
| | 35° | | | | | 0.695 | 0.359 | 0.251 | 0.179 |
| | 40° | | | | | | 0.608 | 0.298 | 0.200 |
| | 45° | | | | | | | 0.518 | 0.238 |
| | 50° | | | | | | | | 0.428 |
| | 0° | 0.592 | 0.511 | 0.441 | 0.378 | 0.323 | 0.273 | 0.228 | 0.189 |
| | 5° | 0.658 | 0.559 | 0.476 | 0.405 | 0.343 | 0.288 | 0.240 | 0.197 |
| | 10° | 0.760 | 0.623 | 0.520 | 0.437 | 0.366 | 0.305 | 0.252 | 0.206 |
| | 15° | 1.129 | 0.723 | 0.581 | 0.478 | 0.395 | 0.325 | 0.267 | 0.216 |
| | 20° | | 1.103 | 0.679 | 0.535 | 0.432 | 0.351 | 0.284 | 0.228 |
| 10° | 25° | | | 1.062 | 0.628 | 0.484 | 0.383 | 0.305 | 0.242 |
| | 30° | | | | 1.005 | 0.571 | 0.430 | 0.334 | 0.260 |
| | 35° | | | | | 0.935 | 0.509 | 0.375 | 0.284 |
| | 40° | | | | | | 0.853 | 0.445 | 0.319 |
| | 45° | | | | | | | 0.763 | 0.380 |
| | 50° | | | | | | | | 0.668 |
| | 0° | 0.690 | 0.611 | 0.540 | 0.476 | 0.419 | 0.366 | 0.317 | 0.273 |
| | 5° | 0.774 | 0.673 | 0.588 | 0.514 | 0.449 | 0.389 | 0.336 | 0.287 |
| | 10° | 0.904 | 0.757 | 0.649 | 0.560 | 0.484 | 0.416 | 0.357 | 0.303 |
| | 15° | 1.372 | 0.889 | 0.731 | 0.618 | 0.526 | 0.448 | 0.380 | 0.321 |
| | 20° | | 1.383 | 0.862 | 0.697 | 0.579 | 0.486 | 0.408 | 0.341 |
| 20° | 25° | | | 1.372 | 0.825 | 0.655 | 0.536 | 0.442 | 0.365 |
| | 30° | | | | 1.341 | 0.778 | 0.606 | 0.487 | 0.395 |
| | 35° | | | | | 1.290 | 0.722 | 0.551 | 0.435 |
| | 40° | | | | | | 1.221 | 0.609 | 0.492 |
| | 45° | | | | | | | 1.136 | 0.590 |
| | 50° | | | | | | | | 1.037 |

(*Continued*)

Table 8.2. (*Continued*)

| $\alpha$ | $\beta$ | $\varphi$ 15° | 20° | 25° | 30° | 35° | 40° | 45° | 50° |
|---|---|---|---|---|---|---|---|---|---|
| | | | | | $\delta = 15°$ | | | | |
| | 0° | 0.458 | 0.371 | 0.298 | 0.237 | 0.186 | 0.142 | 0.106 | 0.076 |
| | 5° | 0.503 | 0.400 | 0.318 | 0.251 | 0.195 | 0.148 | 0.100 | 0.078 |
| | 10° | 0.576 | 0.442 | 0.344 | 0.267 | 0.205 | 0.155 | 0.114 | 0.081 |
| | 15° | 0.850 | 0.509 | 0.380 | 0.289 | 0.219 | 0.163 | 0.119 | 0.084 |
| | 20° | | 0.776 | 0.441 | 0.320 | 0.237 | 0.174 | 0.125 | 0.087 |
| −10° | 25° | | | 0.695 | 0.374 | 0.263 | 0.188 | 0.133 | 0.091 |
| | 30° | | | | 0.607 | 0.308 | 0.209 | 0.143 | 0.097 |
| | 35° | | | | | 0.518 | 0.246 | 0.159 | 0.104 |
| | 40° | | | | | | 0.428 | 0.189 | 0.116 |
| | 45° | | | | | | | 0.341 | 0.137 |
| | 50° | | | | | | | | 0.259 |
| | 0° | 0.405 | 0.314 | 0.240 | 0.180 | 0.132 | 0.093 | 0.062 | 0.038 |
| | 5° | 0.445 | 0.338 | 0.255 | 0.189 | 0.137 | 0.096 | 0.064 | 0.039 |
| | 10° | 0.509 | 0.372 | 0.275 | 0.201 | 0.144 | 0.100 | 0.066 | 0.040 |
| | 15° | 0.763 | 0.429 | 0.303 | 0.216 | 0.152 | 0.104 | 0.068 | 0.041 |
| | 20° | | 0.667 | 0.352 | 0.239 | 0.164 | 0.110 | 0.071 | 0.042 |
| −20° | 25° | | | 0.568 | 0.280 | 0.182 | 0.119 | 0.075 | 0.044 |
| | 30° | | | | 0.470 | 0.214 | 0.131 | 0.080 | 0.046 |
| | 35° | | | | | 0.374 | 0.155 | 0.089 | 0.049 |
| | 40° | | | | | | 0.284 | 0.105 | 0.055 |
| | 45° | | | | | | | 0.203 | 0.065 |
| | 50° | | | | | | | | 0.133 |
| | | | | | $\delta = 20°$ | | | | |
| | 0° | | | 0.357 | 0.297 | 0.245 | 0.199 | 0.160 | 0.125 |
| | 5° | | | 0.384 | 0.317 | 0.259 | 0.209 | 0.166 | 0.130 |
| | 10° | | | 0.419 | 0.340 | 0.275 | 0.220 | 0.174 | 0.135 |
| | 15° | | | 0.467 | 0.371 | 0.295 | 0.234 | 0.183 | 0.140 |
| | 20° | | | 0.547 | 0.414 | 0.322 | 0.251 | 0.193 | 0.147 |
| 0° | 25° | | | 0.874 | 0.487 | 0.360 | 0.273 | 0.207 | 0.155 |
| | 30° | | | | 0.789 | 0.425 | 0.306 | 0.225 | 0.166 |
| | 35° | | | | | 0.714 | 0.362 | 0.252 | 0.180 |
| | 40° | | | | | | 0.625 | 0.300 | 0.202 |
| | 45° | | | | | | | 0.532 | 0.241 |
| | 50° | | | | | | | | 0.440 |

(*Continued*)

Table 8.2. (*Continued*)

| $\alpha$ | $\beta$ | 15° | 20° | 25° | 30° | 35° | 40° | 45° | 50° |
|---|---|---|---|---|---|---|---|---|---|
| | | | | | $\delta = 20°$ | | | | |
| | 0° | | | 0.438 | 0.377 | 0.322 | 0.273 | 0.229 | 0.190 |
| | 5° | | | 0.475 | 0.404 | 0.343 | 0.289 | 0.241 | 0.198 |
| | 10° | | | 0.521 | 0.438 | 0.367 | 0.306 | 0.254 | 0.208 |
| | 15° | | | 0.586 | 0.480 | 0.397 | 0.328 | 0.269 | 0.218 |
| | 20° | | | 0.690 | 0.540 | 0.436 | 0.354 | 0.286 | 0.230 |
| 10° | 25° | | | 1.111 | 0.639 | 0.490 | 0.388 | 0.309 | 0.245 |
| | 30° | | | | 1.051 | 0.582 | 0.437 | 0.338 | 0.264 |
| | 35° | | | | | 0.978 | 0.520 | 0.381 | 0.288 |
| | 40° | | | | | | 0.893 | 0.456 | 0.325 |
| | 45° | | | | | | | 0.799 | 0.389 |
| | 50° | | | | | | | | 0.699 |
| | 0° | | | 0.543 | 0.479 | 0.422 | 0.370 | 0.321 | 0.277 |
| | 5° | | | 0.594 | 0.520 | 0.454 | 0.395 | 0.341 | 0.292 |
| | 10° | | | 0.659 | 0.568 | 0.490 | 0.423 | 0.363 | 0.309 |
| | 15° | | | 0.747 | 0.629 | 0.535 | 0.456 | 0.387 | 0.327 |
| | 20° | | | 0.891 | 0.715 | 0.592 | 0.496 | 0.417 | 0.349 |
| 20° | 25° | | | 1.467 | 0.854 | 0.673 | 0.549 | 0.453 | 0.374 |
| | 30° | | | | 1.434 | 0.807 | 0.624 | 0.501 | 0.406 |
| | 35° | | | | | 1.379 | 0.750 | 0.569 | 0.448 |
| | 40° | | | | | | 1.305 | 0.685 | 0.509 |
| | 45° | | | | | | | 1.214 | 0.615 |
| | 50° | | | | | | | | 1.109 |
| | 0° | | | 0.291 | 0.232 | 0.182 | 0.140 | 0.105 | 0.076 |
| | 5° | | | 0.311 | 0.245 | 0.191 | 0.146 | 0.108 | 0.078 |
| | 10° | | | 0.337 | 0.262 | 0.202 | 0.153 | 0.113 | 0.080 |
| | 15° | | | 0.374 | 0.284 | 0.215 | 0.161 | 0.117 | 0.083 |
| | 20° | | | 0.437 | 0.316 | 0.233 | 0.171 | 0.124 | 0.086 |
| −10° | 25° | | | 0.703 | 0.371 | 0.260 | 0.186 | 0.131 | 0.090 |
| | 30° | | | | 0.614 | 0.306 | 0.207 | 0.142 | 0.096 |
| | 35° | | | | | 0.524 | 0.245 | 0.158 | 0.103 |
| | 40° | | | | | | 0.433 | 0.188 | 0.115 |
| | 45° | | | | | | | 0.344 | 0.137 |
| | 50° | | | | | | | | 0.262 |

(*Continued*)

Table 8.2. (*Continued*)

| $\alpha$ | $\beta$ | $\varphi$ 15° | 20° | 25° | 30° | 35° | 40° | 45° | 50° |
|---|---|---|---|---|---|---|---|---|---|
| | | | | | $\delta = 20°$ | | | | |
| | 0° | | | 0.231 | 0.174 | 0.128 | 0.090 | 0.061 | 0.038 |
| | 5° | | | 0.246 | 0.183 | 0.133 | 0.094 | 0.062 | 0.038 |
| | 10° | | | 0.266 | 0.195 | 0.140 | 0.097 | 0.064 | 0.039 |
| | 15° | | | 0.294 | 0.210 | 0.148 | 0.102 | 0.067 | 0.040 |
| | 20° | | | 0.344 | 0.233 | 0.160 | 0.108 | 0.069 | 0.042 |
| −20° | 25° | | | 0.566 | 0.274 | 0.178 | 0.116 | 0.073 | 0.043 |
| | 30° | | | | 0.468 | 0.210 | 0.129 | 0.079 | 0.045 |
| | 35° | | | | | 0.373 | 0.153 | 0.087 | 0.049 |
| | 40° | | | | | | 0.283 | 0.104 | 0.054 |
| | 45° | | | | | | | 0.202 | 0.064 |
| | 50° | | | | | | | | 0.133 |
| | | | | | $\delta = 25°$ | | | | |
| | 0° | | | | 0.296 | 0.245 | 0.199 | 0.160 | 0.126 |
| | 5° | | | | 0.316 | 0.259 | 0.209 | 0.167 | 0.130 |
| | 10° | | | | 0.340 | 0.275 | 0.221 | 0.175 | 0.136 |
| | 15° | | | | 0.372 | 0.296 | 0.235 | 0.184 | 0.141 |
| | 20° | | | | 0.417 | 0.324 | 0.252 | 0.195 | 0.148 |
| 0° | 25° | | | | 0.494 | 0.363 | 0.275 | 0.209 | 0.157 |
| | 30° | | | | 0.828 | 0.432 | 0.309 | 0.228 | 0.168 |
| | 35° | | | | | 0.741 | 0.368 | 0.256 | 0.183 |
| | 40° | | | | | | 0.647 | 0.306 | 0.205 |
| | 45° | | | | | | | 0.552 | 0.246 |
| | 50° | | | | | | | | 0.456 |
| | 0° | | | | 0.379 | 0.325 | 0.276 | 0.232 | 0.193 |
| | 5° | | | | 0.408 | 0.346 | 0.292 | 0.244 | 0.201 |
| | 10° | | | | 0.443 | 0.371 | 0.311 | 0.258 | 0.211 |
| | 15° | | | | 0.488 | 0.403 | 0.333 | 0.273 | 0.222 |
| | 20° | | | | 0.551 | 0.443 | 0.360 | 0.292 | 0.235 |
| 10° | 25° | | | | 0.658 | 0.502 | 0.396 | 0.315 | 0.250 |
| | 30° | | | | 1.112 | 0.600 | 0.448 | 0.346 | 0.270 |
| | 35° | | | | | 1.034 | 0.537 | 0.392 | 0.295 |
| | 40° | | | | | | 0.944 | 0.471 | 0.335 |
| | 45° | | | | | | | 0.845 | 0.403 |
| | 50° | | | | | | | | 0.739 |

(*Continued*)

Table 8.2. (*Continued*)

| $\alpha$ | $\beta$ | $\varphi$ = 15° | 20° | 25° | 30° | 35° | 40° | 45° | 50° |
|---|---|---|---|---|---|---|---|---|---|
| | | | | | $\delta = 25°$ | | | | |
| | 0° | | | | 0.488 | 0.430 | 0.377 | 0.329 | 0.284 |
| | 5° | | | | 0.530 | 0.463 | 0.403 | 0.349 | 0.300 |
| | 10° | | | | 0.582 | 0.502 | 0.433 | 0.372 | 0.318 |
| | 15° | | | | 0.648 | 0.550 | 0.469 | 0.399 | 0.337 |
| | 20° | | | | 0.740 | 0.612 | 0.512 | 0.430 | 0.360 |
| 20° | 25° | | | | 0.894 | 0.699 | 0.569 | 0.469 | 0.387 |
| | 30° | | | | 1.553 | 0.846 | 0.650 | 0.520 | 0.421 |
| | 35° | | | | | 1.494 | 0.788 | 0.594 | 0.466 |
| | 40° | | | | | | 1.414 | 0.721 | 0.532 |
| | 45° | | | | | | | 1.316 | 0.647 |
| | 50° | | | | | | | | 1.201 |
| | 0° | | | | 0.228 | 0.180 | 0.139 | 0.104 | 0.075 |
| | 5° | | | | 0.242 | 0.189 | 0.145 | 0.108 | 0.078 |
| | 10° | | | | 0.259 | 0.200 | 0.151 | 0.112 | 0.080 |
| | 15° | | | | 0.281 | 0.213 | 0.160 | 0.117 | 0.083 |
| | 20° | | | | 0.314 | 0.232 | 0.170 | 0.123 | 0.086 |
| −10° | 25° | | | | 0.371 | 0.259 | 0.185 | 0.131 | 0.090 |
| | 30° | | | | 0.620 | 0.307 | 0.207 | 0.142 | 0.096 |
| | 35° | | | | | 0.534 | 0.246 | 0.159 | 0.104 |
| | 40° | | | | | | 0.441 | 0.189 | 0.116 |
| | 45° | | | | | | | 0.351 | 0.138 |
| | 50° | | | | | | | | 0.267 |
| | 0° | | | | 0.170 | 0.125 | 0.089 | 0.060 | 0.037 |
| | 5° | | | | 0.179 | 0.131 | 0.092 | 0.061 | 0.038 |
| | 10° | | | | 0.191 | 0.137 | 0.096 | 0.063 | 0.039 |
| | 15° | | | | 0.206 | 0.146 | 0.100 | 0.066 | 0.040 |
| | 20° | | | | 0.229 | 0.157 | 0.106 | 0.069 | 0.041 |
| −20° | 25° | | | | 0.270 | 0.175 | 0.114 | 0.072 | 0.043 |
| | 30° | | | | 0.470 | 0.207 | 0.127 | 0.078 | 0.045 |
| | 35° | | | | | 0.374 | 0.151 | 0.086 | 0.048 |
| | 40° | | | | | | 0.284 | 0.103 | 0.053 |
| | 45° | | | | | | | 0.203 | 0.064 |
| | 50° | | | | | | | | 0.133 |

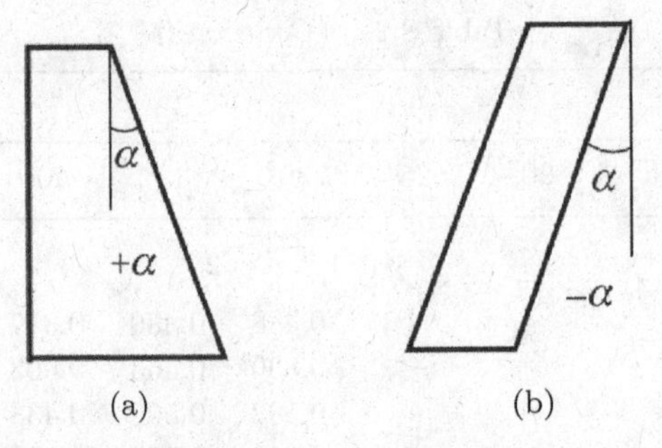


Fig. 8.13. Two conditions for the incline in the retaining wall. (a) Positive value of $\alpha$ and (b) negative value of $\alpha$.

Table 8.3. The evaluation of the friction angle, $\delta$.

| | |
|---|---|
| Smooth wall surface and undrained condition | $\delta = 0^\circ - \varphi/3$ |
| Rough wall surface and undrained condition | $\delta = \varphi/3 - \varphi/2$ |
| Rough wall surface and drained condition | $\delta = \varphi/2 - \varphi$ |

*Note*: $\varphi$ is the friction angle of the soil.

### 8.5.3 *Derivation of the formula for Coulomb's passive earth pressure calculation*

Coulomb's passive earth pressure refers to the wall pushing toward the soil due to the action of the structure, and the wall reaching the limit equilibrium state, which will cause a sliding wedge with an inclined sliding plane. Coulomb's passive earth pressure theory is similar to that of the active earth pressure, only change is in the direction of the wall movement, as shown in Fig. 8.14.

$$E_p = \frac{1}{2}\gamma H^2 \frac{\cos^2(\varphi+\alpha)}{\cos^2\alpha\cos(\alpha-\delta)\left[1-\sqrt{\frac{\sin(\varphi+\delta)\sin(\varphi+\beta)}{\cos(\alpha-\delta)\cos(\alpha-\beta)}}\right]^2}$$

$$= \frac{1}{2}\gamma H^2 K_P, \tag{8.19}$$

$$E_p = \frac{\cos^2(\varphi+\alpha)}{\cos^2\alpha\cos(\alpha-\delta)\left[1-\sqrt{\frac{\sin(\varphi+\delta)\sin(\varphi+\beta)}{\cos(\alpha-\delta)\cos(\alpha-\beta)}}\right]^2}. \tag{8.20}$$

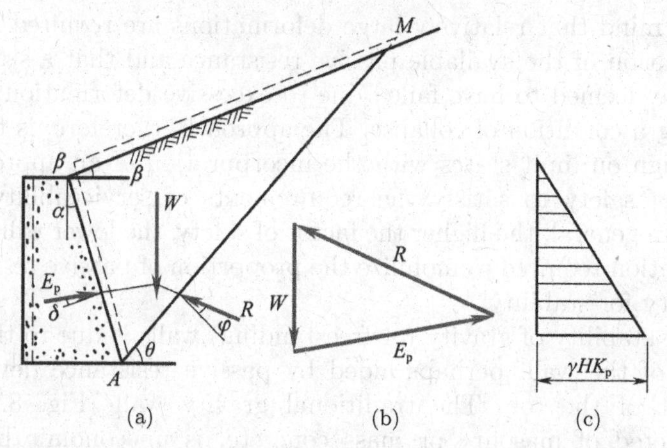


Fig. 8.14. Scheme showing the calculation of Coulomb's active earth pressure. (a) Limit equilibrium of rigid wedge soil; (b) polygon force equilibrium, and (c) gravity of the wedge.

## 8.6 Design of Earth-Retaining Structures

There are two broad categories of retaining structures: (1) gravity, or freestanding walls, in which stability is mainly due to the weight of the structure; (2) embedded walls, in which stability is due to the passive resistance of the soil over the embedded depth and, in most cases, external support. According to the principles of limit state design, and earth-retaining structure must not (a) collapse or suffer major damage, (b) be subject to unacceptable deformations in relation to its location and function, and (c) suffer a minor damage which would necessitate excessive maintenance, rendering it unsightly or reduce its anticipated life. Ultimate limit states are those involving the collapse or instability of the structure as a whole or the failure of one of its components. Serviceability limit states are those involving excessive deformation, leading to a damage or a loss of function. Both ultimate and serviceability limit states must always be considered.

The design of retaining structures has traditionally been based on the specification of a factor of safety in terms of moments, i.e. the ratio of the resisting (or restoring) moment to the disturbing (or overturning) moment. This is known as a lumped factor of safety and is given a value high enough to allow for all the uncertainties in the analytical method and in the values of soil parameters. It must be kept in mind that relatively large deformations are required for the mobilization of the available passive resistance and that a structure could be deemed to have failed due to excessive deformation before reaching a condition of collapse. The approach, therefore, is to base the design on limit states with the incorporation of an appropriate factor of safety to satisfy the requirements of serviceability limit states. In general, the higher the factor of safety, the lower will be the deformation required to mobilize the proportion of passive resistance necessary for stability.

The stability of gravity (or freestanding) walls is due to the self-weight of the wall, perhaps aided by passive resistance developed in front of the toe. The traditional gravity wall (Fig. 8.15(a)), constructed of masonry or mass concrete, is uneconomic because the material is used only for its dead weight. Reinforced concrete cantilever walls (Fig. 8.15(b)) are economical because the backfill itself, acting on the base, is employed to provide most of the required dead weight. Other types of gravity structures include gabion and crib walls (Figs. 8.15(c)–(e)). Gabions are cages of steel mesh, rectangular in plan and elevation, generally filled with cobble-sized particles, with the units being used as the building blocks of a gravity structure. Cribs are open structures assembled from precast concrete or timber members and enclosing coarse-grained fill, with the structure and fill acting as a composite unit to form a gravity wall.

Limit states that must be considered in the design of the retaining wall are as follows:

(1) Overturning of the wall due to instability of the retained soil mass.
(2) Base pressure must not exceed the ultimate bearing capacity of the supporting soil, the maximum base pressure occurring at the toe of the wall because of the eccentricity and inclination of the resultant load.
(3) Sliding between the base of the wall and the underlying soil.
(4) The development of a deep slip surface that envelops the structure as a whole.
(5) Soil and wall deformations that cause adverse effects on the wall itself or on adjacent structures and services.

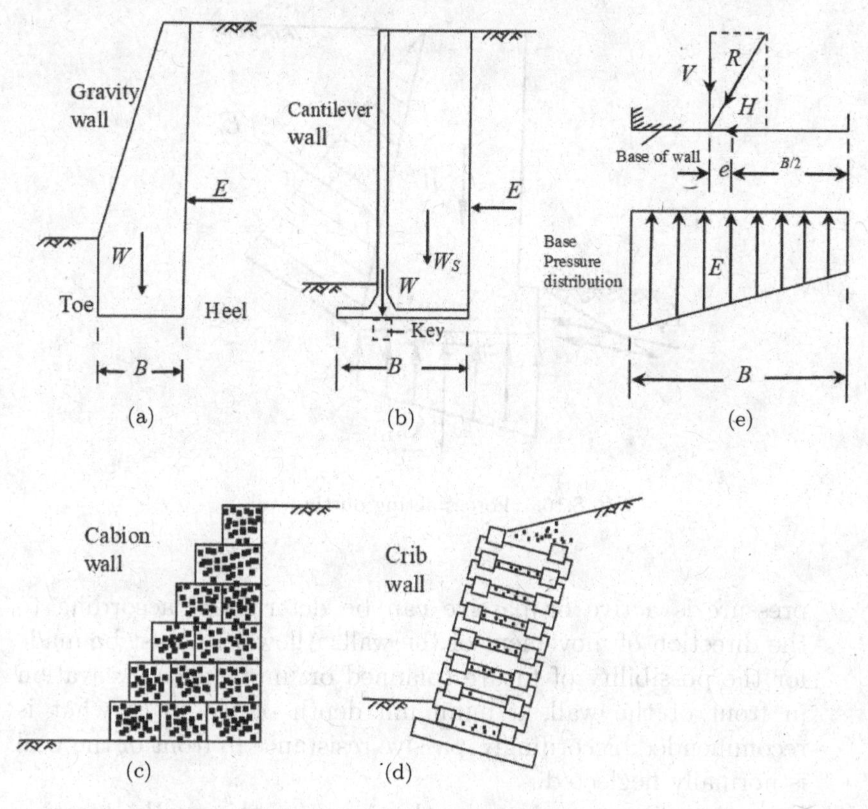


Fig. 8.15. Retaining structures.

(6) Adverse seepage effects, internal erosion, or leakage through the wall: consideration should be given to the consequences of the failure of drainage systems to operate as intended.
(7) Structural failure of any element of the wall or combined soil/structure failure.

The first step in design is to determine all the forces on the wall.

(1) **Self-weight of the wall $W$:** Self-weight of the wall W acting at the center of gravity of the wall. After the type and dimensions of the retaining wall are determined, $W$ is a known value.
(2) **Lateral earth pressure:** Lateral earth pressure is one of the main forces acting on the retaining wall. Whether the earth

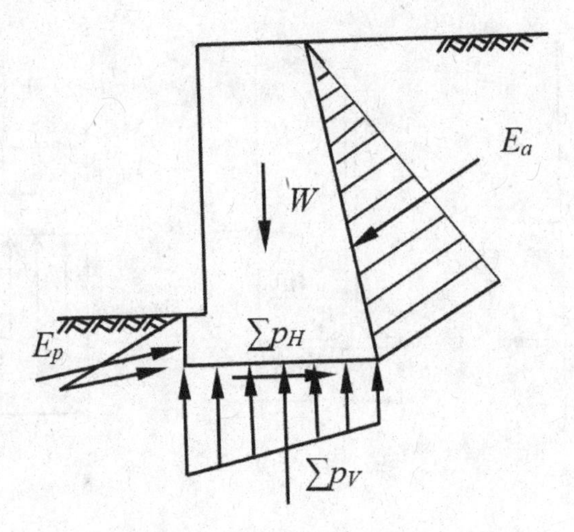


Fig. 8.16. Forces acting on the wall.

pressure is active or passive can be determined according to the direction of movement of the wall. Allowance must be made for the possibility of future (planned or unplanned) excavation in front of the wall, a minimum depth of 0.5 m is what is recommended: accordingly, passive resistance in front of the wall is normally neglected.

(3) **Reaction force acting on the base of the wall:** Reaction force acting on the base of the wall can be divided into horizontal and vertical components. The distribution of the vertical components is assumed to be the same as an eccentrically loaded foundation, i.e. distributed as a trapezoid. The resultant vertical force is represented using $\sum p_V$, which acts at the centroid of the trapezoid. The resultant horizontal force is represented by $\sum p_H$ (Fig. 8.16).

### 8.6.1 *Anti-sliding stability verification*

By dividing the earth pressure $E_a$ into two components,

$$E_{ax} = E_a \cos(\alpha - 90^\circ + \delta), \tag{8.21}$$

$$E_{ay} = E_a \sin(\alpha - 90^\circ + \delta), \tag{8.22}$$

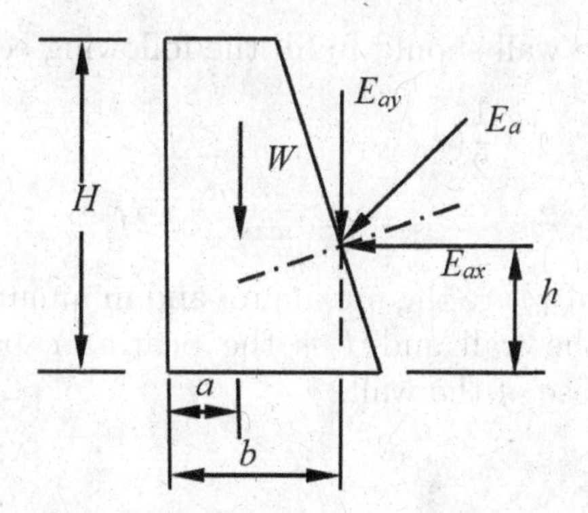


Fig. 8.17. Stability assessment of the wall.

the anti-sliding stability can be calculated using

$$K_s = \frac{(W + E_{ay})\mu}{E_{ax}} \geq 1.3, \tag{8.23}$$

where $K_s$ is the factor of safety of the anti-sliding stability, a minimum of 1.3 of which is specified by the "Code for design of building a foundation". $\mu$ is a coefficient of friction of the base of the wall, which can be determined by tests.

### 8.6.2 *Anti-overturning stability verification*

The anti-overturning stability of the wall can be verified by ensuring that the total resisting moment about the toe exceeds the total overturning moment (Fig. 8.17), i.e.

$$K_t = \frac{Wa + E_{ay}b}{E_{ax}h} \geq 1.5, \tag{8.24}$$

where $K_t$ is the factor of safety of anti-overturning stability, a minimum of 1.5 of which is specified by the "Code for design of building a foundation".

### 8.6.3 *Bearing capacity verification*

The bearing capacity of a retaining wall can be verified using the same method as used in eccentrically loaded foundations. The stress at the base of the wall should fulfill the following equations:

$$\frac{1}{2}(\sigma_{\max} + \sigma_{\min}) \leq f, \tag{8.25}$$

$$\sigma_{\max} \leq 1.2f, \tag{8.26}$$

where $\sigma_{\max}$ and $\sigma_{\min}$ are the maximum and minimum stresses acting on the base of the wall and $f$ is the bearing capacity of the soil underlying the base of the wall.

## Exercises

8.1. For a retaining wall, the underground water table is under the bottom of the wall. The filling soil is sand, the unit weight $\gamma = 17.0\,\text{kN/m}^3$, the height of the wall is 6 m, and the surface of the filling behind the wall is flat, and the value of strength parameters are $\varphi = 37°$ and $c = 0$. Try to calculate the active earth pressure acting on the wall using Rankine's theory. What is the value of earth pressure $E_a$ acting on the wall if the wall is stationary?

8.2. The height of a retaining wall $H = 4.0\,\text{m}$, the rear surface of the wall is vertical and smooth, and the top surface of the filling behind the wall is flat. The filling soil is dry sand, unit weight $\gamma = 18.0\,\text{kN/m}^3$, and the friction angle $\varphi = 36°$. Try to calculate the static earth pressure $E_0$ and the active earth pressure $E_a$ acting on the rear surface of the wall.

8.3. The height of a retaining wall $H = 5.0\,\text{m}$, the width of the top of the wall $b = 1.5\,\text{m}$, and the width of the bottom of the wall $b = 2.5\,\text{m}$. The front surface of the wall is vertical, the rear surface of the wall is inclined, and the friction angle between the rear surface of the wall and the soil $\delta = 20°$. The dip angle $\beta$ of the top surface of the filling behind the wall is 12°. Filling soil is coarse sand, $\gamma = 17.0\,\text{kN/m}^3$ and $\varphi = 30°$. Try to determine the magnitude, direction, and location of the active earth pressure $E_a$ acting on the back surface of the wall.

8.4. The retaining wall is inclined, $\alpha = 20°$, and the dip angle $\beta$ of the top surface of the filling is equal to 10°; the friction angle $\delta$ between the rear surface of the wall and the soil is equal to 15°, the height of the wall is equal to 4.0 m. Filling soil is coarse sand, $\gamma = 20.0\,\text{kN/m}^3$, $c = 0$, and $\varphi = 30°$. Try to determine the magnitude, direction, and location of the active earth pressure $E_a$ acting on the back surface of the wall.

8.5. The height of a retaining wall $H = 10\,\text{m}$, the rear surface of the wall is vertical and smooth, and the surface of filling behind the wall is flat. The upper layer is sand, and the thickness $h_1 = 3.0\,\text{m}$, the unit weight is $\gamma_1 = 18.5\,\text{kN/m}^3$, $\varphi_1 = 30°$; the lower layer is coarse sand, the unit weight is $\gamma_2 = 19.0\,\text{kN/m}^3$, $\varphi_2 = 35°$; the distance of the underground water table from the bottom of the wall is 6.0 m; and the unit weight of the coarse sand underwater is $\gamma_{\text{sat}} = 20.0\,\text{kN/m}^3$. Calculate the magnitude and the location of total active earth pressure ($E_a$) and the pore water pressure ($E_w$) acting on the rear surface of the wall.

## Bibliography


J. B. Burland, D. M. Potts, and N. M. Walsh (1981). The overall stability of free and propped embedded cantilever retaining walls. *Ground Eng.*, 14(5): 28–38.

A. Caquot and J. Kerisel (1966). *Traite de Mecanique des Sols* (Fourth edition). Gauthier-Villars, Paris.

W. C. Huntington (1957). *Earth Pressures and Retaining Walls.* John Wiley and Sons, New York.

T. S. Ingold (1979). The effects of compaction on retaining walls. *Geotechnique*, 29(3): 265–283.

J. Jaky (1944). The coefficient of earth pressure at rest. *J. Soc. Hung. Arch. Eng.*, 7(22): 355–358.

H. Liao (2018). *Soil Mechanics* (Third Edition). Higher Education Press, Beijing.

K. R. Massarsch (1979). Lateral earth pressure in normally consolidated clay. *Proceedings of the Seventh European Conference on Soil Mechanics and Foundation Engineering*, Brighton, England, vol. 2, pp. 245–250.

W. M. J. Rankine (1857). On stability on loose earth. *Phil. Trans. Roy. Soc., Lond.*, Part I: 9–27.

K. Terzaghi and R. B. Peck (1967). *Soil Mechanics in Engineering Practice.* John Wiley and Sons, Inc., New York.

Appendix A

# English–Chinese–Japanese Translation of Frequently used Terminologies

| | | |
|---|---|---|
| **A** | | |
| **active earth pressure** | 主动土压力 | 主働土圧 |
| **additional stress** | 附加应力 | 増加応力 |
| **allowable bearing capacity** | 容许承载力 | 許容支持力 |
| **allowable settlement** | 容许沉降 | 許容沈下 |
| **alluvial fan** | 冲积扇 | 扇状地 |
| **alluvial plain** | 冲积平原 | 沖蹟平野 |
| **angle of internal friction** | 内摩擦角 | 内部摩擦角 |
| **angle of repose** | 休止角 | 安息角 |
| **anisotropy** | 各向异性 | 異方性 |
| **anti-seismic design** | 抗震设计 | 耐震設計 |
| **artesian water** | 承压水 | 被圧水 |
| **axial load** | 轴向荷载 | 軸荷重 |
| **axial symmetry** | 轴对称 | 軸対称 |
| **B** | | |
| **backfill** | 回填土 | 裏込め |
| **back pressure** | 反压力 | 背圧 |
| **bearing capacity factor** | 承载力系数 | 支持力係数 |

| | | |
|---|---|---|
| **bearing capacity of ground** | 地基承载力 | 地盤の支持力 |
| **bearing pile** | 支承桩 | 支持杭 |
| **bulk modulus** | 体积模量 | 体積弾性係数 |

**C**

| | | |
|---|---|---|
| **Cam-Clay model** | 剑桥模型 | カムクレイ・モデル |
| **capillarity** | 毛细作用 | 毛管現象 |
| **capillarity pressure** | 毛细管压力 | 毛管圧力 |
| **capillary water** | 毛细水 | 毛管水 |
| **cast-in-place pile** | 灌注桩 | 現場打ち杭 |
| **circular arc analysis** | 圆弧分析法 | 円形滑り面法 |
| **circular slip surface** | 圆弧滑动面 | 円形滑り面 |
| **clay** | 黏粒/黏土 | 粘土 |
| **coarse gravel** | 粗砾 | 粗れき |
| **coarse sand** | 粗砂 | 粗砂 |
| **coefficient of active earth pressure** | 主动土压力系数 | 主働土圧係数 |
| **coefficient of consolidation** | 固结系数 | 圧密係数 |
| **coefficient of earth pressure at rest** | 静止土压力系数 | 静止土圧係数 |
| **coefficient of lateral pressure** | 侧压力系数 | 側圧係数 |
| **coefficient of passive earth pressure** | 被动土压力系数 | 受働土圧係数 |
| **coefficient of permeability** | 渗透系数 | 透水係数 |
| **coefficient of vertical consolidation** | 竖向固结系数 | 鉛直圧密係数 |
| **cohesion** | 黏聚力 | 粘着力 |
| **cohesive soil** | 黏性土 | 粘性土 |
| **compaction curve** | 击实曲线 | 締め固め曲線 |
| **compaction test** | 击实试验 | 締め固め試験 |
| **compressibility** | 压缩性 | 圧縮性 |
| **compression coefficient** | 压缩系数 | 圧縮係数 |

| | | |
|---|---|---|
| **compression curve** | 压缩曲线 | 圧縮曲線 |
| **compression index** | 压缩指数 | 圧縮指数 |
| **concentrated load** | 集中荷载 | 集中荷重 |
| **confined compression test** | 侧限压缩试验 | 拘束圧縮試験 |
| **consistency** | 稠度 | コンシステンシー |
| **consistency index** | 稠度指数 | コンシステンシー指数 |
| **consolidated drained triaxial test(CD)** | 固结排水三轴试验 | 圧密排水三軸試験 |
| **consolidated undrained triaxial test(CU)** | 固结不排水三轴试验 | 圧密非排水三軸試験 |
| **consolidation settlement** | 固结沉降 | 圧密沈下 |
| **constant head permeability test** | 常水头渗透试验 | 定水位透水試験 |
| **constitutive equation** | 本构方程 | 構成方程式 |
| **constitutive relation** | 本构关系 | 構成式 |
| **controlled-strain triaxial test** | 应变控制式三轴试验 | ひずみ制御三軸試験 |
| **Coulomb's theory of earth pressure** | 库仑土压力理论 | クーロンの土圧論 |
| **creep** | 蠕变 | クリープ |
| **critical hydraulic gradient** | 临界水力梯度 | 限界動水勾配 |
| **critical void ratio** | 临界孔隙比 | 限界間隙比 |
| **crushed stone** | 碎石 | 砕石 |

D

| | | |
|---|---|---|
| **Darcy's law** | 达西定律 | ダルシーの法則 |
| **deep excavation** | 深开挖 | 深い掘削 |
| **deep foundation** | 深基础 | 深い基礎 |
| **deformation** | 变形 | 変形 |
| **degree of consolidation** | 固结度 | 圧密度 |
| **degree of saturation** | 饱和度 | 飽和度 |
| **deviator stress** | 偏应力 | 軸差応力 |
| **dilatancy** | 剪胀性 | ダイレイタンシー |
| **direct shear apparatus** | 直剪仪 | 直接(一面)せん断試験機 |

| | | |
|---|---|---|
| **direct shear test** | 直剪试验 | 直接(一面)せん断 |
| **distributed load** | 分布荷载 | 分布荷重 |
| **disturbed samples** | 扰动土样 | 乱した試料 |
| **drainage condition** | 排水条件 | 排水条件 |
| **dry density** | 干密度 | 乾燥密度 |
| **dry unit weight** | 干重度 | 乾燥単位体積重量 |

**E**

| | | |
|---|---|---|
| **earth dam** | 土坝 | アースダム |
| **earth pressure** | 土压力 | 土圧 |
| **earth pressure at rest** | 静止土压力 | 静止土圧 |
| **earthquake** | 地震 | 地震 |
| **earthquake resistance structure** | 抗震结构 | 耐震構造 |
| **eccentricity** | 偏心距 | 偏心距離 |
| **eccentric load** | 偏心荷载 | 偏心荷重 |
| **effective angle of internal friction** | 有效内摩擦角 | 有効摩擦角 |
| **effective cohesion** | 有效黏聚力 | 有効粘着力 |
| **effective cross section** | 有效截面积 | 有効断面積 |
| **effective diameter** | 有效粒径 | 有効粒径 |
| **effective principal stress** | 有效主应力 | 有効主応力 |
| **effective stress** | 有效应力 | 有効応力 |
| **effective stress path** | 有效应力路径 | 有効応力径路 |
| **effective unit weight** | 有效重度 | 有効単位体積重量 |
| **elastic deformation** | 弹性变形 | 弾性変形量 |
| **elastic limit** | 弹性极限 | 弾性限度 |
| **envelope of Mohr's circles** | 摩尔圆包线 | モールの包絡線 |
| **excavated slope** | 挖边坡 | 切取り斜面 |
| **excess pore water pressure** | 超孔隙水压力 | 過剰間隙水圧 |
| **expansive soil** | 膨胀土 | 膨張性土 |

**F**

| | | |
|---|---|---|
| **factor of compaction** | 压实系数 | 締め固め係数 |
| **failure criteria** | 破坏准则 | 破壊基準 |
| **failure envelope** | 破坏包线 | 破壊包絡線 |
| **falling-head permeability test** | 变水头渗透试验 | 変水位透水試験 |
| **fine sand** | 细砂 | 細砂 |
| **finite element method** | 有限元法 | 有限要素法 |
| **flexible foundation** | 柔性基础 | たわみ性基礎 |
| **flocculent structure** | 絮凝结构 | 綿毛構造 |
| **flow net** | 流网 | 流線網 |
| **flow path** | 流径 | 流動径路 |
| **flow rule** | 流动法则 | 流れ則 |
| **foundation settlement** | 基础沉降 | 基礎沈下 |
| **free groundwater** | 自由地下水 | 自由地下水 |
| **frost heave** | 冻胀 | 凍上 |
| **frozen soil** | 冻土 | 凍土 |

**G**

| | | |
|---|---|---|
| **geological age** | 地质年代 | 地質年代 |
| **geostatic stress** | 自重应力 | 自重応力 |
| **grainage** | 粒度 | 粒度 |
| **grain diameter/particle size** | 粒径 | 粒径 |
| **grain size distribution curve** | 粒径级配曲线 | 粒径加積曲線 |
| **grain size fraction/grain group** | 粒组 | 粒度組成・粒径分布 |
| **gravel** | 砾石 | れき |
| **gravitational water** | 重力水 | 重力水 |
| **gravity retaining wall** | 重力式挡土墙 | 重力式擁壁 |
| **ground** | 地基 | 地盤 |
| **groundwater table** | 地下水位 | 地下水位 |

| | | |
|---|---|---|
| **H** | | |
| **hydraulic conductivity** | 透水性 | 透水性 |
| **hydraulic gradient** | 水力梯度 | 動水勾配 |
| **hydrostatic pressure** | 静水压力 | 静水圧 |
| **I** | | |
| **illite** | 伊利石 | イライト |
| **immediate settlement** | 瞬时沉降 | 即時沈下 |
| ***in situ* testing** | 原位试验 | 原場(原位置) 試験 |
| **instability** | 失稳 | 斜面崩壊 |
| **intact specimen** | 原状土样 | 傷の無い試料 |
| **interlocking** | 咬合作用 | かみ合い |
| **isotropy** | 各向同性 | 等方性 |
| **J** | | |
| **joint** | 节理 | 節理・ジョイント |
| **K** | | |
| **kaolinite** | 高岭石 | カオリナイト |
| **L** | | |
| **laminar flow** | 层流 | 層流 |
| **landslide** | 滑坡 | 地滑り |
| **lateral deformation** | 侧向变形 | 横変形(側方変位) |
| **lateral load** | 侧向荷载 | 横荷重 |
| **layered soil** | 成层土 | 成層土 |
| **line load** | 线荷载 | 線荷重 |
| **linear elastic theory** | 线弹性理论 | 線形弾性理論 |
| **liquid limit** | 液限 | 液性限界 |
| **liquidity index** | 液性指数 | 液性指数 |
| **loading plate test** | 荷载试验 | 載荷試験 |
| **load–settlement curve** | 荷载-沉降曲线 | 荷重-沈下曲線 |
| **local shear failure** | 局部剪切破坏 | 局所せん断破壊 |
| **loess** | 黄土 | レース |

| | | |
|---|---|---|
| **M** | | |
| **maximum dry density** | 最大干密度 | 最大乾燥密度 |
| **mineral composition** | 矿物成分 | 鉱物組成 |
| **mineralogical analysis** | 矿物分析 | 鉱物分析 |
| **modulus of compressibility** | 压缩模量 | 圧縮弾性係数 |
| **Mohr–Coulomb criterion** | 摩尔-库仑准则 | モール・クーロン破壊基準 |
| **Mohr's stress circle** | 摩尔圆 | モールの応力円 |
| **montmorillonite** | 蒙脱石 | モンモリロナイト |
| **N** | | |
| **natural ground** | 天然地基 | 自然地盤 |
| **natural slope** | 天然边坡 | 自然斜面 |
| **natural void ratio** | 天然孔隙比 | 自然間隙比 |
| **natural water content** | 天然含水量 | 自然含水比 |
| **nonlinear analysis** | 非线性分析 | 非線形分析 |
| **nonlinear elastic model** | 非线性弹性模型 | 非線形弾性モデル |
| **non-uniform settlement** | 不均匀沉降 | 不同沈下 |
| **normal stress** | 法向应力 | 垂直応力 |
| **normally consolidated soil** | 正常固结土 | 正規圧密土 |
| **O** | | |
| **oedometer test** | 固结试验 | 圧密試験 |
| **oedometric modulus** | 侧限压缩模量 | 拘束圧縮弾性係数 |
| **one-dimensional consolidation** | 单向固结 | 一次元圧密 |
| **optimum water content** | 最优含水量 | 最適含水比 |
| **organic soil** | 有机质土 | 有機質土 |
| **overconsolidated soil** | 超固结土 | 過圧密土 |
| **overconsolidation ratio** | 超固结比 | 過圧密比 |

**P**

| | | |
|---|---|---|
| **passive earth pressure** | 被动土压力 | 受働土圧 |
| **peak strength** | 峰值强度 | ピーク強度 |
| **peat** | 泥炭 | 泥炭 |
| **permeability** | 渗透性 | 透水性 |
| **physical weathering** | 物理风化 | 物理の風化 |
| **pile foundation** | 桩基 | 杭基 |
| **piping** | 管涌 | パイピング |
| **plane strain** | 平面应变 | 平面ひずみ |
| **plastic deformation** | 塑性变形 | 塑性変形 |
| **plastic limit** | 塑限 | 塑性限界 |
| **plastic zone** | 塑性区 | 塑性域 |
| **plasticity index** | 塑性指数 | 塑性指数 |
| **Poisson's ratio** | 泊松比 | ポアソン比 |
| **pore air pressure** | 孔隙气压力 | 間隙空気圧 |
| **pore pressure coefficient** | 孔隙压力系数 | 間隙圧係数 |
| **pore water pressure** | 孔隙水压力 | 間隙水圧 |
| **porosity** | 孔隙率 | 間隙率 |
| **preconsolidation pressure** | 先期固结压力 | 先行圧密応力 |
| **pressure** | 压力 | 圧力 |
| **principle of effective stress** | 有效应力原理 | 有効応力の原理 |

**Q**

| | | |
|---|---|---|
| **quantity of flow** | 流量 | 流量 |

**R**

| | | |
|---|---|---|
| **Rankine's theory of earth pressure** | 朗肯土压力理论 | ランキンの土圧論 |
| **relative surface area** | 比表面积 | 比表面積 |
| **residual deformation** | 残余变形 | 残留変形 |
| **residual strength** | 残余强度 | 残留強度 |
| **residual stress** | 残余应力 | 残留応力 |
| **retaining wall** | 挡土墙 | 擁壁 |

| | | |
|---|---|---|
| **rheology** | 流变学 | レオロジー・流動学 |
| **rigid foundation** | 刚性基础 | 剛体(剛性)基礎 |

**S**

| | | |
|---|---|---|
| **safety factor** | 安全系数 | 安全率 |
| **sand** | 砂粒/砂 | 砂 |
| **saturated density** | 饱和密度 | 飽和密度 |
| **saturated soil** | 饱和土 | 飽和土 |
| **saturated unit weight** | 饱和重度 | 飽和単位体積重量 |
| **secondary consolidation settlement** | 次固结沉降 | 二次圧密沈下 |
| **seepage flow** | 渗流 | 浸透流 |
| **seepage force** | 渗流力 | 浸透力 |
| **seismic intensity** | 地震烈度 | 震度 |
| **seismic load** | 地震荷载 | 地震荷重 |
| **seismic response** | 地震反应 | 地震応答 |
| **seismic response spectrum** | 地震反应谱 | 地震応答スペクトル |
| **seismic wave** | 地震波 | 地震波 |
| **settlement of ground** | 地基沉降 | 地盤沈下 |
| **settlement** | 沉降 | 沈下 |
| **shallow foundation** | 浅基础 | 浅い基礎 |
| **shear failure** | 剪切破坏 | せん断破壊 |
| **shear modulus** | 剪切模量 | せん断弾性係数 |
| **shear plane $\Delta$, shear surface** | 剪切面 | せん断面 |
| **shear strain** | 切应变 | せん断ひずみ |
| **shear strength** | 抗剪强度 | せん断強さ(強度) |
| **shear stress** | 切应力 | せん断応力 |
| **shear test** | 剪切试验 | せん断試験 |
| **shrinkage limit** | 缩限 | 収縮限界 |
| **silt** | 粉粒/粉土 | シルト |
| **silt fraction** | 粉粒粒组 | シルト分 |
| **silty clay** | 粉质粘土 | シルト質粘土 |
| **silty sand** | 粉砂 | シルト質砂 |

| | | |
|---|---|---|
| **single drainage** | 单面排水 | 片面排水 |
| **slip surface** | 滑裂面 | 滑り面 |
| **slope** | 边坡 | 斜面・のり面 |
| **slope stability** | 边坡稳定性 | 斜面の安定 |
| **soft foundation** | 软弱地基 | 軟弱地盤 |
| **soil mechanics** | 土力学 | 土質力学 |
| **soil particle** | 土粒 | 土粒子 |
| **specific gravity test** | 比重试验 | 比重試験 |
| **stability coefficient** | 稳定系数 | 安定係数 |
| **standard penetration test** | 标准贯入试验 | 標準貫入試験 |
| **static cone penetration test** | 静力触探试验 | 静的円錐貫入試験 |
| **strain hardening** | 应变硬化 | ひずみ硬化 |
| **strain path** | 应变路径 | 応変径路 |
| **strain softening** | 应变软化 | ひずみ軟化 |
| **stress** | 应力 | 応力 |
| **stress analysis** | 应力分析 | 応力分析 |
| **stress distribution** | 应力分布 | 応力分布 |
| **stress history** | 应力历史 | 応力履歴 |
| **stress path** | 应力路径 | 応力径路 |
| **stress–strain curve** | 应力–应变曲线 | 応力–ひずみ曲線 |
| **stress–strain relationship** | 应力–应变关系 | 応力–ひずみ関係 |
| **stress tensor** | 应力张量 | 応力テンソル |
| **strip foundation** | 条形基础 | 帯状基礎 |
| **strip load** | 条形荷载 | 帯状荷重 |
| **submersion** | 浸水 | 浸水 |
| **surface water** | 地表水 | 地表水 |
| **Swedish circle method** | 瑞典圆弧法 | スウェーデン法 |

**T**

| | | |
|---|---|---|
| **Terzaghi's theory of one-dimensional consolidation** | 太沙基一维固结理论 | テルツアーギ一次元圧密モデル |
| **time factor** | 时间因子 | 時間係数 |

| | | |
|---|---|---|
| **total principal stress** | 总主应力 | 全主応力 |
| **total stress approach** | 总应力法 | 全応力解析法 |
| **total stress path** | 总应力路径 | 全応力径路 |
| **triaxial apparatus** | 三轴仪 | 三軸試験機 |
| **triaxial compression test** | 三轴固结试验 | 三軸圧縮試験 |
| **triaxial shear test** | 三轴剪切试验 | 三軸せん断試験 |
| **true triaxial test** | 真三轴试验 | 真の三軸試験 |

**U**

| | | |
|---|---|---|
| **ultimate bearing capacity** | 极限承载力 | 極限支持力 |
| **ultimate load** | 极限荷载 | 極限荷重 |
| **unconfined compression strength** | 无侧限抗压强度 | 一軸圧縮試験 |
| **unconsolidated undrained triaxial test (UU)** | 不固结不排水三轴试验 | 非圧密非排水三軸試験 |
| **unit weight** | 重度 | 単位体積重量 |
| **unsaturated soil** | 非饱和土 | 不飽和土 |

**V**

| | | |
|---|---|---|
| **vane shear test** | 十字板剪切试验 | ベーンせん断試験 |
| **void ratio** | 孔隙比 | 間隙比 |
| **volumetric strain** | 体积应变 | 体積ひずみ |

**W**

| | | |
|---|---|---|
| **water content/moisture content** | 含水量 | 含水比 |
| **weathering** | 风化 | 風化 |

**Y**

| | | |
|---|---|---|
| **yield function** | 屈服函数 | 降伏関数 |
| **yield stress** | 屈服应力 | 降伏応力 |
| **Young's modulus** | 杨氏模量 | ヤング率 |

Appendix B

# Answers to the Exercises

## Chapter 1

1.1 Tips: The definitions of plasticity index and liquidity index and their engineering application.
1.2 Tips: The concept of particle size distribution curve and its engineering application.
1.3 Tips: According to void ratio, relative density, and standard penetration test.
1.4 Tips: The concepts of the maximum dry density and optimum water content.
1.5 $0.972, 1.37\,g/cm^3$
1.6 $1.8\,g/cm^3, 1.61\,g/cm^3, 12\%, 40.5\%, 47.6\%$
1.7 169.6 kg
1.8 Medium dense
1.9 $15.7\,kN/m^3, 19.7\,kN/m^3, 9.9\,kN/m^3, 18.7\,kN/m^3$, 19.3%

## Chapter 2

2.1 Tips: The effect of capillary water on soils and buildings due to movement.
2.2 Tips: The concept of Darcy's Law and its valid condition.
2.3 Tips: According to the valid range of the constant-head and falling-head test as well as the calculation principle the coefficient of permeability.

2.4 Tips: The forming condition of the heaving sand (internal and external causes) and its prevention methods.
2.5 $7.5 \times 10^{-2}$ cm/s
2.6 1.02
2.7 The heaving sand happens.
2.8 5.76 m

## Chapter 3

3.1 0, 35 kPa, 53 kPa, 72 kPa, 93 kPa
3.2 8.24 kPa
3.3 $\sigma_{z=1} = 252.49$ kPa, $\sigma_{z=2} = 140.37$ kPa, $\sigma_{z=3} = 80.34$ kPa, $\sigma_{z=4} = 53.58$ kPa
3.4 19.5%
3.5 Under the point $o$: $\sigma_{z=1} = 124.79$ kPa, $\sigma_{z=2} = 75.4$ kPa, $\sigma_{z=3} = 40.36$ kPa, $\sigma_{z=5} = 18.59$ kPa, $\sigma_{z=6} = 13.13$ kPa. Under the point $A$: $\sigma_{z=1} = 100.24$ kPa, $\sigma_{z=2} = 50.26$ kPa, $\sigma_{z=3} = 31.28$ kPa, $\sigma_{z=4} = 21.35$ kPa, $\sigma_{z=5} = 13.91$ kPa, $\sigma_{z=6} = 10.41$ kPa
3.6 81.3 kPa

## Chapter 4

4.1 $0.9, 0.8, 0.6\,\text{MPa}^{-1}$, 3.2 MPa
4.2 Under-consolidated
4.3 59.3 mm
4.4 156.6 mm, 166 mm
4.5 $6.33 \times 10^{-3}$ cm$^2$/s

## Chapter 5

5.1 (1) 15.1°, 9.8 kPa; (2) No failure
5.2 (1) 31°; (2)150 kPa; (3)206 kPa, 66 kPa
5.3 Failure
5.4 $c = 0$, $\varphi = 41.8°$

## Chapter 6

6.1 125.8 kPa, 146.2 kPa
6.2 (1) $p_{cr} = 82.6$ kPa, $p_{1/4} = 89.7$ kPa, $p_{1/3} = 92.1$ kPa; (2) $p_{cr} = 82.6$ kPa, $p_{1/4} = 85.8$ kPa, $p_{1/3} = 86.9$ kPa
6.3 (1) 236.5 kPa, 94.6 kPa; (2) 269 kPa, 302 kPa

## Chapter 7

7.1 1.88
7.2 0.89
7.3 1.19
7.4 12.5 m
7.5 37°

## Chapter 8

8.1 76.5 kN/m, 122 kN/m
8.2 59.3 kN/m, 37.4 kN/m
8.3 100.5 kN/m, 1.67 m from the bottom of the wall, 31.3° with horizontal direction
8.4 89.6 kN/m, 1.33 m from the bottom of the wall, 35° with horizontal direction
8.5 297.6 kN/m, 3.9 m from the bottom of the wall; 80 kN/m, 1.33 m from the bottom of the wall

# Index


**A**

active earth pressure, 189–190, 196, 214
active earth pressure coefficient, 197
additional stress, 57, 72
additional stress in plane problem, 87
allowable bearing capacity, 144
anti-overturning stability, 219
anti-sliding stability, 219
artificial slopes, 163
at-rest earth pressure, 189–193, 195

**B**

bearing capacity, 219–220
Bishop's method, 178
bulk density, 7
Boussinesq solutions, 74
buoyant density, 7
buoyant unit weight, 176

**C**

capillary phenomena, 33
capillary water zone, 33
Casagrande method, 20
circular slip surface, 173–175
circular slips, 162
coefficient of curvature, 14
coefficient of earth pressure at rest, 62
coefficient of permeability, 38, 41
coefficient of the additional stress, 78
coefficient of uniformity, 14
cohesion, 125
cohesionless soil slope, 166–167, 169
cohesionless soils, 14
cohesive soil, 169
cohesive soil slope, 168–169
compound slips, 163
compression coefficient, 104–105
compression index, 104–105
consolidation, 103
consolidation settlement, 103
constant-head test, 39
contact pressure, 66
contact pressure distribution, 65
Coulomb's earth pressure, 194
Coulomb's law, 167, 171
Coulomb's passive earth pressure, 214
Coulomb's theory, 200
critical edge pressure, 146, 147
critical hydraulic gradient, 50
critical safety height, 182
critical void ratio, 133, 135

**D**

Darcy's law, 35
degree of consolidation, 117
degree of saturation, 8
dilatancy, 132
dip angle, 201

discharge velocity, 37
downward seepage, 48
dry density, 24

E

earth pressure, 188–191
earth pressure distribution, 191
effective cohesion, 176
effective friction angle, 176, 193
effective normal reaction force, 179
effective normal stress, 63
effective overburden pressure, 57
effective reaction force, 179
effective size, 13
effective stress principle, 62
effective unit weight, 168
embedded walls, 188, 215
engineering geology, 162
equilibrium state, 188
equipotential line, 47

F

factor of safety, 165, 180, 215, 219
failure envelope, 126
falling-head test, 40
Fellenius, 169–170
flocculent fabric, 12
flow in the horizontal direction, 42
flow in the vertical direction, 44
flow line, 47
flow net, 46
foundation, 56
friction angle, 124, 167, 177, 201
friction force, 202
fully saturated soil, 7

G

general shear failure, 144
geotechnical engineering, 56, 162
geotechnical structures, 2
gravity wall, 188, 216
gravity, or freestanding walls, 215
ground, 56

H

homogeneous, 166
homogeneous soil slope, 172
honeycomb fabric, 11
horizontal geostatic stress, 62
horizontal uniform load, 92
hydraulic gradient, 35, 168

I

interpolation method, 203

J

Jaky's method, 193

L

laminar flow, 34
large deformations, 216
lateral earth pressure, 192, 217
lateral forces, 178
lateral pressure, 188
layerwise summation method,
  108–109
limit equilibrium method, 144
liquefaction, 165
liquidity index, 19, 23
load distribution, 66
local shear failure, 145

M

maximum density, 17
maximum principal stress, 195, 197
method of sedimentation, 14
method of sieving, 13
minimum density, 17
minimum principal stress, 195–197
Mohr circle, 126
Mohr theory, 194
Mohr–Coulomb envelope, 137
Mohr–Coulomb failure criterion, 128

N

natural slopes, 163
non-circular slip surface, 170

non-circular slips, 162
normal stress, 58

**O**

oedometric modulus, 104, 106
one-way eccentric load, 69
overconsolidated clays, 193
overconsolidation ratio, 138, 193

**P**

particle size analysis, 12
particle size distribution curve, 14
passive earth pressure, 189
passive earth pressure coefficient, 198
passive pressure, 190
permeability, 32
Petersen, 170
Petterson, 169
phase relationships, 5
plastic collapse, 188
plastic limit, 18
plasticity index, 19, 22
Poisson's ratio, 193
polygon force equilibrium principle, 200–202
pore pressure, 180
pore stress coefficient, 180
pore water pressure, 63, 131, 164, 175, 178, 181
pore water stress, 176
porosity, 8
principle of superposition, 78
punching shear failure, 145

**R**

Rankine's earth pressure theory, 194, 200
Rankine's theory, 193, 196, 203
reaction force, 218
rectangular foundation, 78
relative density, 17
residual strength, 132, 137
Reynolds number, 36
rigid block, 201
rigid body, 171
Rubbing method, 19

**S**

saturation weight, 168, 176
seepage, 32
seepage flow, 181
seepage flow force, 164
seepage force, 47, 167–168, 176, 181
seismic load, 165
semi-infinite soil, 194
settlement, 102–103
shear failure, 124
shear resistance, 124
shear shrinkage, 132
shear strength, 124–126, 135, 162, 164, 167, 169, 171, 174
shear strength parameters, 126
shear stress, 58, 162, 164
shearing resistance force, 168
shrinkage limit, 18
single grain fabric, 11
slicing method, 170
sliding force, 165–166
sliding wedge soil mass, 200
slip resistance, 173
slip resistant force, 165
slope angle, 164, 201
soil classification, 26
soil compaction, 23
soil compaction curve, 25
soil composition, 3
soil deformation, 56
soil fabric, 11
soil mass, 189–190
soil mechanics, 2, 162
soil parameters, 215
soil skeleton, 62
soil stability, 56
soil stratum, 163
specific gravity, 7

stability number, 181
stable slope angle, 182
standard penetration test, 18
static equilibrium, 190
static liquefaction, 49
strain hardening, 132
strain softening, 132
stress components, 59
stress increment, 73
stress paths, 130
stress state of soil, 57
strip foundation, 88
supporting soil, 216
Swedish slice method, 172

**T**

tangential force, 166
tangential reaction force, 179
Taylor, 169, 181
total normal stress, 63
traditional gravity wall, 216
translational slips, 163
triangularly distributed load, 83
turbulent flow, 35
two-way eccentric load, 71

**U**

ultimate bearing capacity, 144, 150
unit weight, 7, 164, 173
upward seepage, 49

**V**

vertical centric load, 68
vertical concentrated load, 73
vertical geostatic stress, 60
vertical line load, 87
vertical uniform load, 75
void ratio, 7, 15, 107

**W**

water content, 6

Printed in the United States
by Baker & Taylor Publisher Services